ADDITIONAL SKILL AND DRILL MANUAL

JAMES J. BALL
Indiana State University

BASIC COLLEGE MATHEMATICS

SEVENTH EDITION

Margaret L. Lial
American River College

Stanley A. Salzman
American River College

Diana L. Hestwood
Minneapolis Community and Technical College

PEARSON

Addison
Wesley

Boston San Francisco New York
London Toronto Sydney Tokyo Singapore Madrid
Mexico City Munich Paris Cape Town Hong Kong Montreal

Reproduced by Pearson Addison-Wesley from electronic files supplied by the author.

Copyright © 2006 Pearson Education, Inc.
Publishing as Pearson Addison-Wesley, 75 Arlington Street, Boston, MA 02116.

ISBN   0-321-33168-0

        8 BRR 08

CONTENTS

ADDITIONAL EXERCISES

Chapter 1

WHOLE NUMBERS

1.1 Reading and Writing Whole Numbers

Objective 1   **Identify whole numbers.**

Indicate whether each number is a whole number or not a whole number.

1. 48   2. 1.2   3. $7\frac{1}{2}$ 4. 0

5. 142   6. 357   7. 12.47   8. $6\frac{3}{4}$

9. $\frac{5}{11}$ 10. 10,029   11. 3.14149   12. 0.216

Objective 2   **Give the place value of a digit.**

Fill in the digit for the given place value in each of the following whole numbers.

13. 9841	thousands	tens
14. 4336	thousands	ones
15. 25,016	ten-thousands	hundreds
16. 86,331	ten-thousands	ones
17. 5,813,207	millions	thousands
18. 2,800,439,012	billions	millions

Fill in the number for the given period in each of the following whole numbers.

19. 29,176	thousands	ones	
20. 75,229,301	millions	thousands	ones
21. 70,000,603,214	billions	millions	thousands	ones
22. 300,459,200,005	billions	millions	thousands	ones

Objective 3   **Write a number in words or digits.**

Rewrite the following numbers in words.

23. 8714   24. 39,015   25. 834,768

26. 2,015,102   27. 96,543,228   28. 499,324,518

Rewrite each of the following numbers using digits.

29. Four thousand, one hundred twenty-seven

30. Twenty-nine thousand, five hundred sixteen

31. Six hundred eight-five million, two hundred fifty-nine

32. Three hundred million, seventy-five thousand, two

Rewrite the number from the following sentence using digits.

33. A bottle of a certain vaccine will give seven thousand, two hundred ten injections.

34. Every year, nine hundred seventy-two thousand, four hundred thirty people visit a certain historical area.

35. A supermarket has fifteen thousand three hundred thirteen different items for sale.

36. The population of a large city is six million, two hundred five thousand.

| Objective 4 | Read a table.

Use the table to find each of the following and write the number in digits.

37. The number of calories burned by a 130-pound adult in 30 minutes of cycling.

38. The activity which will burn at least 300 calories when performed by a 123-pound adult.

39. The number of calories burned by a 143-pound adult in 30 minutes of jumping rope.

40. The weight of an adult who will burn 189 calories in 30 minutes of walking.

Weight of Exerciser	123 lbs	130 lbs	143 lbs
	Calories burned in 30 minutes		
Cycling	168	177	195
Running	324	342	375
Jumping Rope	273	288	315
Walking	162	171	189

Source: Fitness magazine

1.1 Mixed Exercises

Indicate whether each number is a whole number or not a whole number.

41. 57   42. 2.067   43. $\dfrac{17}{20}$ 44. 410

Rewrite the following numbers in words.

45. 6243   46. 904   47. 16,201

Rewrite the following numbers in digits.

48. Two thousand, three hundred twenty-one

49. Nine hundred thousand, four hundred fifty-six

50. Seven million, six thousand, twelve

Give the place value of the digit 7 in each number.

51. 27,632   52. 248,071   53. 764,592

WHOLE NUMBERS

1.2 Adding Whole Numbers

Objective 1   Add two single-digit numbers.

Add.

1. 3 + 5	2. 6 + 4	3. 3 + 9	4. 6 + 6	5. 7 + 6
6. 4 + 7	7. 8 + 8	8. 8 + 5	9. 2 + 9	10. 3 + 7
11. 7 + 7	12. 6 + 5	13. 2 + 8	14. 9 + 6	15. 7 + 8
16. 7 + 9	17. 4 + 5	18. 8 + 3	19. 6 + 8	20. 9 + 8

Objective 2   Add more than two numbers.

Add.

21.	22.	23.	24.	25.
7	6	9	2	6
2	7	3	3	5
5	2	2	9	3
3	1	1	7	9
+ 8	+ 5	+ 4	+ 4	+ 2

26.	27.	28.	29.	30.
9	2	8	4	9
7	5	5	6	7
2	3	7	7	8
5	2	3	2	3
+ 6	1	2	5	6
	+ 4	+ 1	+ 9	+ 4

31.	32.	33.	34.	35.
3	4	7	6	3
4	6	8	2	8
9	3	3	8	9
2	1	2	7	2
7	9	1	9	4
+ 6	+ 8	+ 5	+ 4	+ 2

36.	6	37.	6	38.	7	39.	8	40.	2
	4		1		2		1		7
	9		7		9		9		8
	8		2		4		6		4
	4		9		7		3		7
	+ 3		+ 4		+ 6		+ 7		+ 1

Objective 3   Add when carrying is not required.

Add.

41.   11
       + 58

42.   42
       + 57

43.   56
       + 22

44.   37
       + 51

45.   82
       + 15

46.   142
       + 336

47.   421
       + 567

48.   836
       + 142

49.   781
       + 114

50.   304
       + 572

51.   42,305
       + 10,574

52.   33,412
       + 65,265

53.   46,314
       + 23,462

54.   86,305
       + 12,672

55.   38,204
       + 20,392

56.   5705
       121
       + 4163

57.   6310
       2252
       + 1337

58.   135
       253
       + 7410

59.   314
       2121
       + 6263

60.   45,158
       20,340
       +   2401

Objective 4   Add the following numbers by carrying as necessary.

61.   42
       + 39

62.   83
       + 29

63.   124
       + 94

64.   356
       + 278

65.   872
       + 459

66.   563
       + 478

67.   384
       + 219

68.   9258
       + 763

69.   43,648
       + 5932

70.   72,150
       + 39,394

71.   38,204
       + 21,807

72.   9481
       + 7686

73. 7439
    + 8376

74. 6739
     5085
    + 7327

75. 3057
       69
     6078
    + 642

76. 7033
      809
     2532
    + 41

77. 3197
      420
      638
       67
    + 3774

78. 524
      96
    8828
     703
    + 29

79. 946
    1023
      85
     672
    + 701

80. 2432
      702
        3
       28
    + 2707

Objective 5 **Solve application problems with carrying.**

Using the map below, find the shortest distance between the following cities.

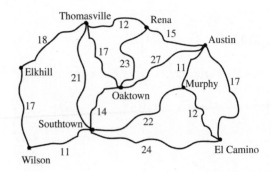

81. Murphy and Thomasville

82. Elkhill and Austin

83. Wilson and Austin

84. El Camino and Thomasville

Solve the following application problems, using addition.

85. A box of applies costs $27 and a box of peaches $23. Find the total cost for a box of each.

86. Kevin Levy has 52 nickels, 37 dimes, and 119 quarters. How many coins does he have altogether?

87. There were 325 women and 365 men at the school craft fair. How many people were at the fair?

88. The theater sold 276 adult tickets, and 349 child tickets. How many tickets were sold altogether?

89. At a charity bazaar, a church has a total of 1873 books for sale, while a lodge has 3358 books for sale. How many books are for sale?

90. A plane is flying at an altitude of 5830 feet. It then increases its altitude by 7384 feet. Find its new altitude.

Find the perimeter or total distance around each of the following figures.

91.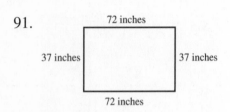
72 inches
37 inches   37 inches
72 inches

92.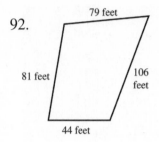
79 feet
81 feet   106 feet
44 feet

93.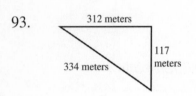
312 meters
117 meters
334 meters

94.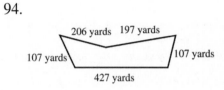
206 yards   197 yards
107 yards   107 yards
427 yards

Objective 6   **Check the answer in addition.**

Check the following additions. If an answer is incorrect, give the correct answer.

95.	96.	97.	98.
67	34	398	82
48	76	214	47
+ 83	+ 81	+ 65	+ 976
198	191	577	905

99.	100.	101.	102.
3217	326	73	341
1408	9014	9815	1270
932	761	390	27
+ 7255	+ 8329	+ 7002	+ 9488
12,812	18,330	16,270	11,126

103.	104.	105.	106.
723	881	3028	395
681	790	335	8740
29	30	2914	32
412	5015	688	315
+ 103	+ 769	+ 1647	+ 294
1947	7485	8612	9876

107.	7148	108.	3576	109.	72	110.	683
	6850		4241		38		3863
	92		3007		5735		22
	6003		86		764		951
	+ 17		+ 912		+ 16		+ 66
	20,110		10,822				5785

1.2 Mixed Exercises

111. 9 + 4   112. 3 + 8   113. 7 + 8   114.   6
                                                                                7
      5
   + 8

115.   9   116.   9   117.   42   118.   624
    2   2   + 57   + 372
 + 7   8
        + 7

119.   49   120.   5928
 + 76   3647
       + 9266

Solve the following application problems.

121.   John earned $672 one week and $659 the next week. Find the total amount earned in the two weeks.

122.   Gena has 14 nickels, 28 dimes, and 31 quarters. How many coins does she have?

Check the following additions. If an answer is incorrect, give the correct answer.

123.	738	124.	31	125.	27	126.	68
	804		6271		319		5894
	19		8		5		7203
	2911		5862		8720		15
	+ 305		+ 57		+ 419		+ 6819
	4777		12,229		9390		19,999

WHOLE NUMBERS

1.3 Subtracting Whole Numbers

Objective 1   Change addition problems to subtraction and subtraction problems to addition.

Write two subtraction problems for each addition problem.

1. $3 + 4 = 7$

2. $6 + 9 = 15$

3. $8 + 5 = 13$

4. $15 + 8 = 23$

5. $17 + 9 = 26$

6. $19 + 13 = 32$

7. $37 + 25 = 62$

8. $89 + 23 = 112$

9. $47 + 83 = 130$

10. $47 + 35 = 82$

11. $149 + 38 = 187$

12. $253 + 59 = 312$

13. $478 + 239 = 717$

14. $476 + 538 = 1014$

Write an addition problem for each subtraction problem.

15. $1211 - 426 = 785$

16. $787 - 183 = 604$

17. $204 - 87 = 117$

18. $313 - 47 = 266$

19. $5917 - 2196 = 3721$

20. $5094 - 113 = 4981$

Objective 2   Identify the minuend, subtrahend, and difference.

Identify the minuend, subtrahend, and difference in each of the following subtraction problems.

21. $5 - 3 = 2$

22. $7 - 2 = 5$

23. $22 - 7 = 15$

24. $35 - 9 = 24$

25. $36 - 27 = 9$

26. $98 - 36 = 62$

27. $18 - 12 = 6$

28. $47 - 29 = 18$

29. $187 - 36 = 151$

30. $236 - 142 = 94$

Objective 3   Subtract when no borrowing is needed.

Subtract.

31. $\begin{array}{r} 95 \\ -\ 64 \\ \hline \end{array}$

32. $\begin{array}{r} 86 \\ -\ 25 \\ \hline \end{array}$

33. $\begin{array}{r} 63 \\ -\ 41 \\ \hline \end{array}$

34. 58
    − 38
	35. 76
	    − 55
		36. 157
		    − 132

37. 563
    − 412
	38. 563
	    − 242
		39. 923
		    − 613

40. 5573
    − 422
	41. 942
	    − 732
		42. 8417
		    − 4206

43. 4376
    − 2154
	44. 3986
	    − 1475
		45. 8539
		    − 2527

46. 25,493 − 13,271		

Objective 4   **Check answers.**

Check the following subtractions. If an answer is not correct, give the correct answer.

47. 75
    − 32
    ‾‾‾
    43
	48. 63
	    − 29
	    ‾‾‾
	    36
		49. 68
		    − 57
		    ‾‾‾
		    11

50. 47
    − 39
    ‾‾‾
    16
	51. 192
	    − 39
	    ‾‾‾
	    167
		52. 625
		    − 376
		    ‾‾‾
		    351

53. 8739
    − 3892
    ‾‾‾‾
    4847
	54. 9328
	    − 7142
	    ‾‾‾‾
	    2186
		55. 4847
		    − 3768
		    ‾‾‾‾
		    1121

56. 5763
    − 2783
    ‾‾‾‾
    3980
	57. 3857
	    − 2135
	    ‾‾‾‾
	    1722
		58. 4210
		    − 1650
		    ‾‾‾‾
		    2660

59. 7000
    − 3984
    ‾‾‾‾
    3016
	60. 2001
	    − 563
	    ‾‾‾‾
	    1438
		61. 7122
		    − 6914
		    ‾‾‾‾
		    208

62.	82,314	63.	75,000	64.	31,146
	− 19,514		− 67,109		− 7312
	72,800		7891		23,834

65.	82,004	66.	64,397
	− 3917		− 27,008
	79,193		37,389

Objective 5   Subtract by borrowing.

67.	42	68.	92	69.	72
	− 35		− 47		− 43

70.	87	71.	54	72.	94
	− 48		− 29		− 48

73.	613	74.	927	75.	437
	− 421		− 729		− 259

76.	724	77.	4687	78.	33,728
	− 657		− 2769		− 7829

79.	86,372	80.	70	81.	60
	− 29,485		− 27		− 44

82.	300
	− 57

Objective 6   Solve application problems with subtraction.

Solve the following application problems.

83. A kennel has 72 dogs. It sells 40. How many dogs are left?

84. A Girl Scout has 52 boxes of cookies to sell. If she sells 27 boxes, how many boxes will she have left?

85. An airplane is carrying 234 passengers. When it lands in Atlanta, 139 passengers get off the plane. How many passengers are then left on the plane?

86. Nathaniel Best has $553 in his checking account. He writes a check for $134. How much is then left in the account?

87. Barbara Wolters drove her truck 741 miles, while Arnold Parker drove his car 396 miles. How many more miles did Wolters drive?

88. On Sunday, 7342 people went to a football game, while on Monday, 9138 people went. How many more people went on Monday?

89. In last fall's election, 3645 people voted. In this fall's election, 2511 people voted. How many more people voted in last fall's election?

90. One bid for painting a house was $2134. A second bid was $1954. How much would be saved using the second bid?

91. Sally Tanner had $22,143 withheld from her paycheck last year for income tax. She actually owes only $16,959 in tax. What refund should she receive?

92. A retired couple gets a social security payment of $439 per month. Recently benefits were increased to $515 per month. How much more money does the couple now get?

93. The Conrads now pay $439 per month for rent. If they rent a larger apartment, the payment will be $702 per month. How much extra will they pay each month?

94. A truck now goes 342 miles on a tank of gas. After a tune-up, the same truck will go 405 miles. How many additional miles will it go after the tune-up?

95. On Friday, 11, 594 people visited the Eastridge Amusement Park, while 14,352 people visited the park on Saturday. How many more people visited the park on Saturday?

1.3 Mixed Exercises

Write two subtraction problems for each addition problem.

96. $26 + 31 = 57$

97. $627 + 43 = 670$

Write an addition problem for each subtraction problem.

98. $49 - 27 = 22$

99. $165 - 37 = 128$

Identify the minuend, subtrahend, and difference in each of the following subtraction problems.

100. $62 - 17 = 45$

101. $424 - 79 = 345$

Subtract.

102.	45,136	103.	58,932	104.	57,364	105.	86,804
	$-\,31{,}015$		$-\,32{,}701$		$-\,6253$		$-\,3702$

106. 108
       − 69

107. 400
       − 192

108. 37,000
       − 22,338

109. 44,000
       − 28,352

Solve the following application problems.

110. At People's Bank, Marc Lukas can earn $1538 per year in interest, while Farmer's Bank would pay him $1643 interest. How much additional interest would he earn at the second bank?

111. The Folkes now make a house payment of $931 per month. If they buy a larger house, the payment will be $1215 per month. How much more will they pay each month for the larger house?

112. Last Year, 574 athletes competed in a district track meet at Johnson College. This year, 498 athletes competed. How many fewer athletes competed this year than last?

WHOLE NUMBERS

1.4 Multiplying Whole Numbers

Objective 1   **Know the parts of a multiplication problem.**

Identify the factors and the product in each multiplication problem.

1. $5 \times 3 = 15$ 2. $4 \times 7 = 28$ 3. $8 \cdot 4 = 32$ 4. $5(2) = 10$

5. $(9)(8) = 72$ 6. $14 \times 1 = 14$ 7. $9 \cdot 0 = 0$ 8. $13 \cdot 3 = 39$

9. $56 = 7 \cdot 8$ 10. $108 = 9 \times 12$ 11. $17 \cdot 5 = 85$ 12. $(1)(5) = 5$

Objective 2   **Do Chain multiplication.**

Multiply.

13. $4 \times 4 \times 2$ 14. $3 \times 4 \times 7$ 15. $4 \cdot 2 \cdot 1$ 16. $7 \cdot 4 \cdot 0$

17. $0 \cdot 3 \cdot 8$ 18. $(7)(1)(5)$ 19. $(6)(4)(8)$ 20. $5 \cdot 5 \cdot 2$

21. $2 \cdot 6 \cdot 10$ 22. $3 \times 4 \times 5$

Objective 3   **Multiply by single-digit numbers.**

Multiply.

23. 5×8 24. 0×9 25. $8 \cdot 6$ 26. $7 \cdot 7$

27. $(3)(7)$ 28. $8 \cdot 7$ 29. $6 \cdot 9$ 30. 8×7

31. 4×9 32. $(4)(3)$ 33. $9(7)$ 34. 2×11

35. $5 \cdot 12$ 36. $1 \cdot 18$

$$
\begin{array}{cccc}
37. \quad 54 & 38. \quad 42 \\
\underline{\times\ 4} & \underline{\times\ 6}
\end{array}
$$

$$
\begin{array}{cccc}
39. \quad 76 & 40. \quad 163 & 41. \quad 297 & 42. \quad 862 \\
\underline{\times\ 3} & \underline{\times\ 5} & \underline{\times\ 8} & \underline{\times\ 7}
\end{array}
$$

$$
\begin{array}{cccc}
43. \quad 753 & 44. \quad 398 & 45. \quad 523 & 46. \quad 405 \\
\underline{\times\ 6} & \underline{\times\ 5} & \underline{\times\ 9} & \underline{\times\ 7}
\end{array}
$$

47.   408
     × 7

48.   2004
     × 7

49.   3008
     × 3

50.   21,010
     × 4

51.   30,009
     × 6

52.   7359
     × 2

53.   6270
     × 5

54.   93,105
     × 6

55.   38,471
     × 3

56.   31,763
     × 9

Objective 4 Use multiplication shortcuts for numbers ending in 0s.

Multiply.

57. 82×10

58. 91×100

59. 429×10

60. 439×1000

61. (852)(30)

62. 103·400

63. 42×200

64. 873·600

65. 209(500)

66. 3005×2000

67. 409·7000

68. (6073)(80)

69. 8234×2000

70. 8701×500

71. 387·20,000

72. 500×40

73. 820·500

74. 47,000·6000

75. 30·700

76. 8000·600

Objective 5 Multiply by numbers having more than one digit.

Multiply.

77.   46
     × 21

78.   47
     × 32

79.   27
     × 32

80.   76
     × 39

81.   98
     × 76

82.   81
     × 29

83.   644
     × 19

84.   508
     × 23

85.   701
     × 38

86.   409
     × 27

87.   4031
     × 48

88.   8341
     × 59

89.   5249
     × 63

90.   7165
     × 53

91.   7008
     × 39

92.   8621
     × 131

Objective 6 Solve application problems with multiplication.

Solve the following application problems.

93. A fabric store has 16 bolts of silk. Each bolt contains 35 yards of silk. How many yards of silk does the fabric store have in all?

94. On a recent trip the Jensen family drove 45 miles per hour on the average. They drove 22 hours altogether. How many miles did they drive altogether?

95. Marisa Taylor saves $38 out of every pay check. Last year she received 24 pay checks. How much did she save?

96. Heinen's Supermarket received a shipment of 28 cartons of canned vegetables. There were 24 cans in each carton. How many cans were there altogether?

Find the total cost of each of the following.

97. 18 chairs at $42 per chair

98. 24 soccer balls at $16 per ball

99. 47 watches at $29 per watch

100. 83 ice machines at $435 per machine

101. 175 baseball bats at $16 per bat

102. 512 boxes of chalk at $19 per box

1.4 Mixed Exercises

Identify the factors and the product in each multiplication.

103. $51 = (3)(17)$

104. $39 = 13(3)$

105. $0 \times 18 = 0$

106. $30 = 30(1)$

Multiply.

107. $5 \times 4 \times 3$

108. $(2)(0)(8)$

109. $4 \cdot 7 \cdot 5$

110. 2×17

111. $\begin{array}{r} 362 \\ \times\ 4 \\ \hline \end{array}$

112. $\begin{array}{r} 20,400 \\ \times\ \ \ \ 2 \\ \hline \end{array}$

113. 57×100

114. 712×10

115. 6248×1000

116. $\begin{array}{r} 9532 \\ \times\ 203 \\ \hline \end{array}$

117. $\begin{array}{r} 5109 \\ \times\ 273 \\ \hline \end{array}$

118. $\begin{array}{r} 7391 \\ \times\ 4312 \\ \hline \end{array}$

119. $\begin{array}{r} 8042 \\ \times\ 3409 \\ \hline \end{array}$

Find the total cost of each of the following.

120. 278 tires at $51 per tire

121. 178 baseball hats at $9 per hat

122. 79 clocks at $198 per clock

123. 167 volley balls at $23 per ball

WHOLE NUMBERS

1.5 Dividing Whole Numbers

Objective 1 Write division problems in three ways.

Write each division problem using two other symbols.

1. $15 \div 3 = 5$

2. $\dfrac{18}{6} = 3$

3. $3\overline{)27}$ with quotient 9

4. $\dfrac{39}{13} = 3$

5. $8\overline{)64}$ with quotient 8

6. $\dfrac{50}{25} = 2$

7. $\dfrac{28}{7} = 4$

8. $16\overline{)32}$ with quotient 2

9. $\dfrac{42}{7} = 6$

10. $36 \div 12 = 3$

11. $\dfrac{0}{8} = 0$

12. $24\overline{)0}$ with quotient 0

13. $12\overline{)72}$ with quotient 6

14. $\dfrac{100}{25} = 4$

Objective 2 Identify the parts of a division problem.

Identify the dividend, divisor, and quotient.

15. $27 \div 9 = 3$

16. $63 \div 7 = 9$

17. $5\overline{)30}$ with quotient 6

18. $14\overline{)42}$ with quotient 3

19. $\dfrac{38}{19} = 2$

20. $\dfrac{65}{13} = 5$

21. $28 \div 4 = 7$

22. $18 \div 9 = 2$

23. $52 \div 4 = 13$

24. $7\overline{)35}$ with quotient 5

25. $11\overline{)132}$ with quotient 12

26. $\dfrac{44}{11} = 4$

27. $\dfrac{63}{9} = 7$

28. $39 \div 13 = 3$

Objective 3 Divide 0 by a number.

Divide, whenever possible.

29. $0 \div 15$

30. $\dfrac{0}{6}$

31. $12\overline{)0}$

32. $5\overline{)0}$

33. $0 \div 25$
34. $\dfrac{0}{14}$
35. $0 \div 3$
36. $0 \div 10$

Objective 4 **Recognize that a number cannot be divided by 0.**

Divide, whenever possible.

37. $\dfrac{7}{0}$
38. $0\overline{)72}$
39. $9 \div 0$
40. $0\overline{)31}$

41. $\dfrac{100}{0}$
42. $\dfrac{1}{0}$

Objective 5 **Divide a number by itself.**

Divide.

43. $13 \div 13$
44. $5\overline{)5}$
45. $\dfrac{12}{12}$
46. $18 \div 18$

47. $17\overline{)17}$
48. $9 \div 9$
49. $\dfrac{8}{8}$
50. $200 \div 200$

51. $14\overline{)14}$
52. $\dfrac{1}{1}$

Objective 6 **Divide a number by 1.**

Divide.

53. $17 \div 1$
54. $\dfrac{128}{1}$
55. $1\overline{)38}$
56. $9 \div 1$

57. $\dfrac{27}{1}$
58. $1\overline{)12}$

Objective 7 **Use short division.**

59. $6\overline{)72}$
60. $2\overline{)84}$
61. $39 \div 3$
62. $74 \div 2$

63. $\dfrac{186}{3}$
64. $435 \div 3$
65. $\dfrac{575}{5}$
66. $3\overline{)438}$

67. $\dfrac{512}{3}$
68. $724 \div 5$
69. $6\overline{)247}$
70. $8\overline{)1135}$

71. $843 \div 7$

72. $4\overline{)819}$

73. $\dfrac{651}{9}$

74. $984 \div 6$

Objective 8 Check the answer to a division problem.

Check each answer. If an answer is incorrect, give the correct answer.

75. $5\overline{)135}$ with quotient 28

76. $3\overline{)3150}$ with quotient 1050

77. $6\overline{)9137}$ with quotient 1522 R4

78. $2915 \div 8 = 364$

79. $3852 \div 4 = 963$

80. $\dfrac{11,980}{3} = 3993$ R 1

81. $\dfrac{52,696}{8} = 6587$

82. $\dfrac{8621}{3} = 2873$ R 2

83. $46,650 \div 7 = 6664$ R2

84. $9\overline{)64,198}$ with quotient 7132 R 4

85. $4\overline{)118,315}$ with quotient $29,578$ R 3

86. $7\overline{)40,698}$ with quotient 4814

87. $\dfrac{34,176}{7} = 4882$ R 2

88. $\dfrac{126,104}{8} = 14,763$

89. $3\overline{)587,254}$ with quotient $195,751$ R 1

90. $\dfrac{40,097}{8} = 512$ R 1

91. $\dfrac{18,150}{3} = 650$

92. $20,351 \div 6 = 3391$ R 5

93. $3200 \div 9 = 354$ R 6

94. $9\overline{)30,508}$ with quotient 3389 R 8

Objective 9 Use tests for divisibility.

*Put a ✓ mark in the blank if the number at the left is divisible by the number at the top.
Put an X in the blank if the number is not divisible by the number at the top.*

Check each answer. If an answer is incorrect, give the correct answer.

	2	3	5	10
95. 50	___	___	___	___
96. 36	___	___	___	___
97. 285	___	___	___	___
98. 492	___	___	___	___

		2	3	5	10
99.	375	___	___	___	___
100.	897	___	___	___	___
101.	723	___	___	___	___
102.	908	___	___	___	___
103.	5100	___	___	___	___
104.	1734	___	___	___	___
105.	6205	___	___	___	___
106.	7610	___	___	___	___
107.	2000	___	___	___	___
108.	2583	___	___	___	___
109.	13,302	___	___	___	___
110.	24,179	___	___	___	___
111.	32,175	___	___	___	___
112.	641,238	___	___	___	___
113.	452,650	___	___	___	___
114.	523,894	___	___	___	___

1.5 Mixed Exercises

Write each division problem using two other symbols.

115. $42 \div 6 = 7$ 116. $15\overline{)45}$ with quotient 3 117. $\dfrac{128}{8} = 16$

Identify the dividend, divisor, and quotient.

118. $9\overline{)45}$ with quotient 5 119. $72 \div 9 = 8$ 120. $\dfrac{27}{9} = 3$

Divide, whenever possible.

121. $\dfrac{12}{0}$ 122. $\dfrac{0}{8}$ 123. $0\overline{)216}$

Divide.

124. $16 \div 16$ 125. $9 \div 9$ 126. $47\overline{)47}$

127. $\dfrac{308}{5}$ 128. $2601 \div 7$ 129. $6\overline{)4007}$

130. $600 \div 9$

Use a divisibility test to determine if each statement is true of false.

131.   171 is divisible by 3.   132.   2560 is divisible by 5.

133.   98,648 is divisible by 2.   134.   8714 is divisible by 3.

WHOLE NUMBERS

1.6 Long Division

Objective 1   Do long division.

Divide using long division. Check each answer.

1. $32\overline{)2624}$

2. $29\overline{)9396}$

3. $42\overline{)3234}$

4. $23\overline{)1587}$

5. $53\overline{)5406}$

6. $37\overline{)4215}$

7. $89\overline{)7649}$

8. $94\overline{)29,047}$

9. $71\overline{)412,794}$

10. $86\overline{)8,473,758}$

11. $205\overline{)6,680,335}$

12. $327\overline{)98,413,712}$

13. $657\overline{)429,700}$

14. $732\overline{)4,268,292}$

Objective 2   Divide numbers in 0 by numbers ending in 0.

Divide.

15. $30\overline{)270}$

16. $80\overline{)560}$

17. $500\overline{)4500}$

18. $700\overline{)4900}$

19. $2000\overline{)12,000}$

20. $6000\overline{)72,000}$

21. $50\overline{)800}$

22. $400\overline{)6000}$

23. $230\overline{)16,100}$

24. $910\overline{)38,220}$

25. $210\overline{)16,800}$

26. $750\overline{)25,500}$

27. $500\overline{)42,000}$

28. $1200\overline{)960,000}$

Objective 3   Check answers.

Check each answer. If an answer is incorrect, give the correct answer.

29. $19\overline{)3299}^{\,173}$ R 12

30. $37\overline{)3235}^{\,87}$ R 16

31. $89\overline{)5790}^{\,65}$ R 5

32. $74\overline{)25,621}^{\,346}$ R 18

33. $103\overline{)4658}^{\,44}$ R 22

34. $205\overline{)47,538}^{\,231}$ R 183

35. $318\overline{)94,207}$ quotient 297 R 79   36. $428\overline{)196,883}$ quotient 400 R 30   37. $537\overline{)431,042}$ quotient 802 R 368

38. $614\overline{)152,923}$ quotient 249 R 37

1.6 Mixed Exercises

Divide using long division. Check each answer.

39. $81\overline{)4293}$ 40. $39\overline{)8073}$ 41. $56\overline{)9314}$ 42. $28\overline{)177,919}$

43. $103\overline{)672,796}$ 44. $523\overline{)2,098,276}$ 45. $70\overline{)5600}$ 46. $800\overline{)10,400}$

47. $7000\overline{)189,000}$ 48. $900\overline{)207,000}$ 49. $5200\overline{)46,800}$ 50. $330\overline{)20,130}$

WHOLE NUMBERS

1.7 Rounding Whole Numbers

Objective 1 **Locate the place to which a number is to be rounded.**

Locate the place to which the number is rounded by underlining the appropriate digit.

1.	853	Nearest ten		2.	1037	Nearest hundred
3.	4712	Nearest thousand		4.	645,371	Nearest ten-thousand
5.	4,316,214	Nearest hundred-thousand		6.	39,943,712	Nearest million
7.	643,519	Nearest hundred		8.	81,243	Nearest ten-thousand
9.	257,301	Nearest ten		10.	2,781,421	Nearest thousand

Objective 2 **Round numbers.**

Round as shown.

11. 7863 to the nearest hundred

12. 449 to the nearest ten

13. 1382 to the nearest ten

14. 4938 to the nearest ten

15. 814 to the nearest ten

16. 18,211 to the nearest hundred

17. 32,576 to the nearest ten

18. 9348 to the nearest hundred

19. 53,595 to the nearest hundred

20. 14,703 to the nearest hundred

21. 8398 to the nearest hundred

22. 41,099 to the nearest hundred

23. 16,668 to the nearest hundred

24. 5312 to the nearest thousand

25. 3842 to the nearest thousand

26. 51,803 to the nearest thousand

Objective 3   **Round numbers to estimate.**

Estimate the following answers by rounding to the nearest ten. Then find the exact answers.

27.	37	28.	19	29.	69	30.	88
	24		87		− 42		− 52
	58		35				
	+ 91		+ 20				

Estimate the following answers by rounding to the nearest hundred. Then find the exact answer.

31.	276	32.	419	33.	971	34.	815
	312		188		− 382		− 678
	174		324				
	+ 936		+ 194				

35.	912	36.	876
	× 784		× 141

Objective 4   **Use front end rounding to estimate an answer.**

Estimate the following answers by using front end rounding. Then find the exact answer.

37.	571	38.	731	39.	872	40.	313
	42		31		− 39		− 49
	215		709				
	+ 2452		+ 78				

41.	980	42.	437
	× 37		× 29

1.7 Mixed Exercises

Locate the place to which the number is rounded by underlining the appropriate digit.

43. 4799   Nearest hundred   44. 782,563   Nearest ten-thousand

45. 28,963,521   Nearest million

Round as indicated.

46. 54,289 to the nearest thousand

47. 476,943 to the nearest ten-thousand

48. 576,295 to the nearest hundred-thousand

49. 14,823,307 to the nearest million

Estimate the following answers by rounding to the nearest hundred. Then find the exact answer.

50.	761	51.	298	52.	691	53.	462
	× 584		× 401		× 751		× 139

Estimate the following answers by using front end rounding. Then find the exact answer.

54.	3210	55.	4710	56.	725	57.	521
	− 856		− 348		× 66		× 81

WHOLE NUMBERS

1.8 Exponents, Roots, and Order of Operations

Objective 1 Identify an exponent and a base.

Identify the exponent and the base. Simplify each expression

1. 7^2

2. 4^2

3. 9^2

4. 3^3

5. 1^5

6. 10^4

7. 2^7

8. 8^3

Objective 2 Find the square root of a number.

Find each square root.

9. $\sqrt{4}$

10. $\sqrt{9}$

11. $\sqrt{16}$

12. $\sqrt{121}$

13. $\sqrt{64}$

14. $\sqrt{100}$

15. $\sqrt{49}$

16. $\sqrt{169}$

Complete each blank.

17. $18^2 =$ _____ so $\sqrt{\rule{1cm}{0pt}}\, = 18$

18. $14^2 =$ _____ so $\sqrt{196} =$ _____

19. $50^2 =$ _____ so $\sqrt{\rule{1cm}{0pt}}\, = 50$

20. $17^2 =$ _____ so $\sqrt{289} =$ _____

21. $23^2 =$ _____ so $\sqrt{\rule{1cm}{0pt}}\, = 23$

22. $25^2 =$ _____ so $\sqrt{\rule{1cm}{0pt}}\, = 25$

23. $15^2 =$ _____ so $\sqrt{\rule{1cm}{0pt}}\, = 15$

24. $36^2 =$ _____ so $\sqrt{1296} =$ _____

25. $20^2 =$ _____ so $\sqrt{400} =$ _____

26. $52^2 =$ _____ so $\sqrt{\rule{1cm}{0pt}}\, = 52$

Objective 3 Use the order of operations.

Work each problem by using the order of operations.

27. $6^2 + 5 - 2$

28. $2^4 + 3 \cdot 4 - 5$

29. $6 \cdot 5 - 5 \div 0$

30. $8 \cdot 9 \div 12$

31. $9 \cdot 7 - 3 \cdot 12$

32. $7 - 25 \div 5$

33. $6 \cdot 3^2 + 0 \div 6$

34. $8 \cdot 5 - 12 \div (2 \cdot 3 - 6)$

35. $4 \cdot 3 + 8 \cdot 5 - 7$

36. $4 \cdot (9 - 7) + 3 \cdot 8$

37. $2^3 \cdot 3^2 + 5(3) \div 5$

38. $6 \cdot \sqrt{144} - 6 \cdot 8$

39. $42 \div 6 + 3 \cdot \sqrt{49}$

40. $2 \cdot \sqrt{121} - 2 \div \sqrt{4} + (14 - 2 \cdot 7) \div 4$

1.8 Mixed Exercises

Identify the exponent and the base. Simplify each expression.

41. 5^3

42. 6^3

43. 3^5

Find each square root.

44. $\sqrt{400}$

45. $\sqrt{1}$

46. $\sqrt{49}$

Use the order of operations.

47. $25 \div 5 \cdot 3 \cdot 9 \div (14 - 11)$

48. $4 \cdot 7 \div \sqrt{49} - 4 \div 1 + (0 + 8)$

49. $3^2 \cdot 2 + 2^3 \cdot 5 - \sqrt{81} \div 3 \cdot 4$

50. $6 \cdot \sqrt{25} \cdot \sqrt{100} \div 3 - \sqrt{4} - 3^2$

51. $2 \cdot 3 - \sqrt{16} + 2 \cdot 3^2 - \sqrt{100} \div \sqrt{4}$

52. $3^2 \cdot \sqrt{36} \div \sqrt{81} \div 3 + 2 \cdot 3 - 2$

WHOLE NUMBERS

1.9 Reading Pictographs, Bar Graphs, and Line Graphs

Objective 1 Read and understand a pictograph.

Use the pictograph to answer the questions.

State Sales Tax

Georgia	$ $ $ $
Utah	$ $ $ $ $
Idaho	$ $ $ $ $
Texas	$ $ $ $ $ [
Minnesota	$ $ $ $ $ [$

$ = 1% sales tax

Source: Federation of Tax Administrators

1. Which state shown in the pictograph charges the least sales tax?

2. Which state has a sales tax of 5%?

3. According to the pictograph, which state has the greatest sales tax?

4. Which state has a sales tax of 4%?

Objective 2 Read and understand a bar graph.

Use the bar graph, which shows the amount of blood donated by the different departments of a company, to answer the questions.

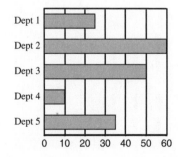

5. How many pints of blood were donated by Department 3?

6. How many more pints of blood were donated by Department 5 than Department 4?

7. Which Department donated the fewest pints of blood?

8. How many pints of blood were donated by Department 1?

Objective 3 **Read and understand a line graph.**

Use the line graph to answer the questions.

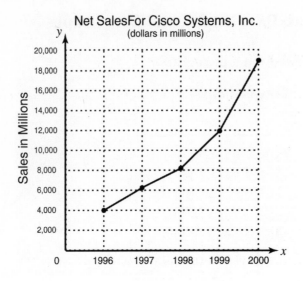

9.   What trend or patterns is shown in the graph?

10. Approximately what were the net sales in 1996?

11. Which year had the largest increase over the previous year?

12. Which year had the highest net sales?

1.9 Mixed Exercises

Use the pictograph to answer the questions.

School Language Clubs Membership

13. Which language club has the largest number of members?

14. How many more female students are there than male students in the French club?

15. What is the total number of members in the Chinese club?

16. Which language club has the least number of members?

Use the double bar graph, which shows the enrollment by gender in each class at a small college, to answer the questions.

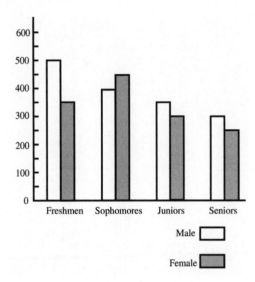

17. How many more male freshmen are there than female seniors?

18. Find the total number of students enrolled.

19. How many more sophomores are there than juniors?

20. Which class has the greatest difference between male students and female students?

Use the comparison line graph, which shows the annual sales for two different stores for each of the past few years, to answer the questions.

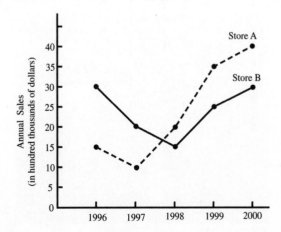

21. In which years did the sales of store A exceed the sales of store B?

22. Which year showed the least difference between the sales of store A and the sales of store B?

23. Which year showed the greatest difference between the sales of store A and the sales of store B?

24. What was the difference in annual sales between store A and store B in 1996?

WHOLE NUMBERS

1.10 Solving Application Problems

Objective 1 **Find indicator words in application problems.**

Write the operation that is indicated by the following words.

1. plus

2. decreased by

3. twice

4. sum of

5. product

6. less than

7. quotient

8. difference

9. total

10. increase of

11. more than

12. per

Objective 2 **Solve application problems.**

Solve each of the following application problems.

13. Andy's Auto Supply just raised the price of a rebuilt engine to $1985. If this is $163 more than the old price, find the old price.

14. Tim Rhinehart, coordinator of Toys for Tots, has collected 2548 toys. If his group wants to give the same number of toys to each of 637 children, how many toys will each child receive?

15. If 843 movie tickets are sold per day, how many tickets will be sold in a 5-day period.

16. If profits of $642,000 are divided evenly among a firm's 1000 employees, how much will each employee receive?

17. Dana Burgess owes $2840 plus $168 interest on his credit union loan. If he wishes to pay the loan in full, how much must he pay?

18. To qualify for a real estate loan at Uptown Bank, a borrower must have a monthly income of at least 4 times the monthly payment. What minimum monthly income must a borrower have to qualify for a monthly payment of $725?

19. On a ship, 192 meters of rope were used to drop anchor. If the rope is 241 meters long, how many meters are not used?

20. Naturalists report that 69 salmon are passing a fish ladder each hour. At this rate, how many salmon are passing the ladder in a 12-hour period?

21. Lori Knight knows that her car gets 36 miles per gallon in town. How many miles can she travel on 26 gallons?

22. A car loan of $12,672 is to be paid off in 36 months. What will the monthly payment be?

23. Diana Ditka spent $286 on tuition, $137 on books, and $32 on supplies. If this money is withdrawn form her checking account, which had a balance of $723, what is her new balance?

24. The Whole Earth Candle Works used 375 pounds of wax, 11 pounds of pigment, 3 pounds of scent, 7 pounds of sparkle grains, and 1 pound of wick material. What is the total weight of the candles made from these materials?

Objective 3 **Estimate an answer.**

Estimate an answer for each problem by using front end rounding. Then find the exact answer.

25. A bus traveled 605 miles at 55 miles per hour. How long did the trip take?

26. A police car traveled 96 miles at 48 miles per hour. How long did it take?

27. The total cost for 23 baseball uniforms is $1817. Find the cost of each uniform.

28. A person borrows $47,000, and pays interest of $541. Find the total amount that must be repaid.

29. A small car goes 648 miles on 18 gallons of gas. How many miles per gallon does the car get?

30. Amanda Raymond owes $5520 on a loan. Find her monthly payment if the loan is paid off in 48 months.

31. Two sisters share a legal bill of $1903. One sister pays $954 toward the bill. How much must the other sister pay?

32. Teisha Jordan can assemble 38 toasters on one hour. How many hours would it take her to assemble 4066 toasters?

33. The total receipts at a concert were $191,800. Each ticket cost $28. How many people attended the concert?

34. A new car costs $11,350 before a trade-in. The car can be paid off in 36 monthly payments of $209 each after the trade-in. Find the amount of the trade-in.

35. A biology class found 14 deer in one area, 158 in another, and 417 in a third. How many deer did the class find?

36. A Seiko quartz watch has a crystal that vibrates 32,768 times in one second. How many times will this crystal vibrate in 30 seconds?

37. Blue Bird leader, Barbara Walton estimates that each of her Blue Birds will eat 2 cookies while she and her assistant, Lana Meehan, will eat 3 cookies each. If she expects 15 Blue Birds and her assistant at the meeting, how many cookies will she need?

38. Edward Biondi has $3117 in his checking account. If he pays $340 for tires, $725 for equipment repairs, and $198 for fuel and oil, find the balance remaining in his account.

39. Cheryl Brown can type 12 forms per hours. How many forms can she type in 7 hours?

1.10 Mixed Exercises

Write the operation that is indicated by the following words.

40. subtracted from   41. divided by   42. goes into

43. product   44. more than   45. twice

46. divided equally   47. per

Solve each of the following application problems.

48. A truck weights 8950 pounds when empty. After being loaded with firewood, it weighs 17,180 pounds. What is the weight of the firewood?

49. How many 3-inch strips of leather can be cut from a piece of leather 1 foot wide? (Hint: 1 foot = 12 inches.)

50. If there are 43,560 square feet in an acre, how many square feet are there in 5 acres?

51. Ski Mart offers a set of skis at a sale price of $219. If the sale price gives a savings of $56 off the original price, what is the original price of the skis?

52. The number of gallons of water polluted each day in an industrial area is 219,530. How many gallons are polluted each year? (Use a 365-day year.)

53. Travel Rent-A-Car owns 365 compact cars, 438 full-sized cars, 125 luxury cars, and 83 vans and trucks. How many vehicles does it have in all?

54. Rodney Guess owns 55 acres of land which he leases to an alfalfa farmer for $150 per acre per year. If property taxes are $28 per acre per year, find the total amount he has left after taxes are paid.

55. A room measures 18 feet by 12 feet. If carpeting costs $23 per square yard, find the total cost for carpeting the room. (Hint: one square yard = 3 feet × 3 feet.)

Estimate an answer for each problem by using front end rounding. Then find the exact answer.

56. If 560 stamps are divided evenly among 16 collectors, how many stamps will each receive?

57. There are 325 words on one page. Find the number of pages needed for 28,925 words.

58. Liz Skinner has $3712 in her checking account. After writing a check of $887 for tuition and parking fees, how much remains in her account?

59. The Top Hat Grille finds that it needs five pounds of hamburger to make 35 servings of chili. How many pounds of hamburger are needed to make 182 servings of chili?

60. Jerri Taft's vending machine company had 325 machines on hand at the beginning of the month. At different times during the month, machines were distributed to new locations: 37 machines were taken at one time, then 24 machines, and then 81 machines. During the same month additional machines were returned: 16 machines were returned at one time, then 39 machines, and then 110 machines. How many machines were on hand at the end of the month?

Chapter 2

MULTIPLYING AND DIVIDING FRACTIONS

2.1 Basics of Fractions

Objective 1   Use fractions to show which part of a whole is shaded.

Write the fraction that represents the shaded area.

1.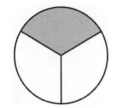

2.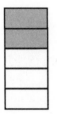

3.

4.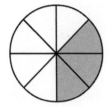

5.

6.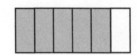
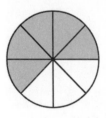

7.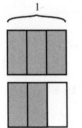

8.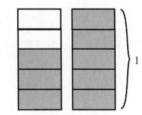

9.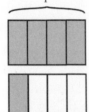

10.

11.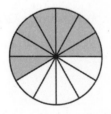

12.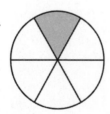

Objective 2   Identify the numerator and denominator.

Identify the numerator and the denominator.

13. $\dfrac{4}{3}$ 14. $\dfrac{1}{2}$ 15. $\dfrac{2}{5}$ 16. $\dfrac{9}{23}$ 17. $\dfrac{8}{11}$ 18. $\dfrac{11}{8}$

19. $\dfrac{112}{5}$ 20. $\dfrac{19}{50}$ 21. $\dfrac{7}{15}$ 22. $\dfrac{19}{8}$ 23. $\dfrac{98}{13}$ 24. $\dfrac{157}{12}$

Objective 3   **Identify proper and improper fractions.**

Write whether each fraction is proper or improper.

25. $\dfrac{9}{7}$ 26. $\dfrac{5}{12}$ 27. $\dfrac{7}{15}$ 28. $\dfrac{17}{11}$

29. $\dfrac{4}{19}$ 30. $\dfrac{1}{4}$ 31. $\dfrac{11}{7}$ 32. $\dfrac{18}{18}$

33. $\dfrac{5}{4}$ 34. $\dfrac{10}{10}$ 35. $\dfrac{7}{12}$ 36. $\dfrac{14}{13}$

37. $\dfrac{2}{2}$ 38. $\dfrac{11}{4}$ 39. $\dfrac{3}{4}$ 40. $\dfrac{21}{17}$

2.1 Mixed Exercises

Write the fraction that represents the shaded area.

41. 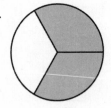   42.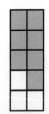

43.    44.

Identify the numerator and denominator.

45. $\dfrac{14}{195}$ 46. $\dfrac{83}{85}$ 47. $\dfrac{42}{23}$ 48. $\dfrac{0}{16}$

Write whether each fraction is **proper** *or* **improper.**

49. $\dfrac{13}{12}$ 50. $\dfrac{9}{10}$ 51. $\dfrac{17}{17}$ 52. $\dfrac{2}{3}$

MULTIPLYING AND DIVIDING FRACTIONS

2.2 Mixed Numbers

Objective 1   **Identify mixed numbers.**

List the mixed numbers in each group.

1. $2\frac{1}{2}, \frac{3}{5}, 1\frac{1}{6}, \frac{3}{4}$

2. $\frac{3}{8}, 5\frac{2}{3}, \frac{7}{4}, 3\frac{1}{2}$

3. $\frac{8}{7}, \frac{10}{10}, \frac{2}{3}, \frac{0}{5}$

Objective 2   **Write mixed numbers as improper fractions.**

Write each mixed number as an improper fraction.

4. $2\frac{7}{8}$

5. $1\frac{5}{6}$

6. $2\frac{4}{5}$

7. $5\frac{4}{7}$

8. $1\frac{3}{4}$

9. $6\frac{1}{4}$

10. $4\frac{2}{3}$

11. $7\frac{1}{2}$

12. $2\frac{7}{11}$

13. $5\frac{3}{7}$

14. $6\frac{2}{3}$

15. $8\frac{7}{9}$

Objective 3   **Write improper fractions as mixed numbers.**

Write each improper fraction as a mixed number.

16. $\frac{11}{2}$

17. $\frac{8}{5}$

18. $\frac{9}{8}$

19. $\frac{33}{10}$

20. $\frac{14}{9}$

21. $\frac{20}{7}$

22. $\frac{29}{9}$

23. $\frac{26}{7}$

24. $\frac{21}{5}$

25. $\frac{41}{9}$

26. $\frac{25}{9}$

27. $\frac{29}{4}$

2.2 Mixed Exercises

List the mixed numbers.

28. $\frac{9}{9}, 3\frac{1}{2}, 10\frac{1}{3}, \frac{8}{2}, \frac{7}{9}$

29. $\frac{6}{3}, 4\frac{3}{4}, \frac{10}{10}, \frac{1}{3}, \frac{0}{8}$

Write each mixed number as an improper fraction.

30. $11\frac{1}{3}$ 31. $4\frac{5}{8}$ 32. $5\frac{3}{8}$ 33. $8\frac{3}{5}$

34. $4\frac{3}{7}$ 35. $7\frac{1}{9}$ 36. $13\frac{3}{9}$ 37. $22\frac{8}{11}$

Write each improper fraction as a mixed number.

38. $\dfrac{58}{5}$ 39. $\dfrac{27}{7}$ 40. $\dfrac{47}{10}$ 41. $\dfrac{56}{13}$

42. $\dfrac{92}{3}$ 43. $\dfrac{211}{11}$ 44. $\dfrac{749}{17}$ 45. $\dfrac{2573}{11}$

MULTIPLYING AND DIVIDING FRACTIONS

2.3 Factors

Objective 1   **Find factors of a number.**

Find all the factors of each number.

1. 7	2. 12	3. 49	4. 15
5. 10	6. 36	7. 25	8. 24
9. 18	10. 30	11. 72	12. 28

Objective 2   **Identify primes.**

Write whether each number is **prime, composite,** *or* **neither.**

13. 1	14. 5	15. 11	16. 15
17. 24	18. 45	19. 2	20. 31
21. 29	22. 38	23. 52	24. 46
25. 81	26. 61	27. 93	28. 85

Objective 3   **Find prime factorizations.**

Find the prime factorization of each number. Write the answer with exponents when repeated factors appear.

29. 12	30. 22	31. 15	32. 27
33. 28	34. 42	35. 32	36. 24
37. 63	38. 100	39. 72	40. 56

2.3 Mixed Exercises

Find all the factors of each number.

41. 40	42. 66

Write whether each number is **prime, composite,** *or* **neither.**

43. 65	44. 39	45. 67	46. 51

Find the prime factorization of each number. Write the answer with exponents when repeated factors appear.

47. 70   48. 108   49. 85   50. 800

51. 240   52. 160   53. 450   54. 171

MULTIPLYING AND DIVIDING FRACTIONS

2.4 Writing a Fraction in Lowest Terms

Objective 1 Tell whether a fraction is written in lowest terms.

Write whether or not each fraction is in lowest terms.

1. $\frac{4}{12}$ 2. $\frac{3}{7}$ 3. $\frac{12}{18}$ 4. $\frac{13}{17}$

5. $\frac{9}{12}$ 6. $\frac{27}{30}$ 7. $\frac{7}{19}$ 8. $\frac{12}{28}$

9. $\frac{4}{8}$ 10. $\frac{4}{9}$ 11. $\frac{18}{21}$ 12. $\frac{3}{39}$

Objective 2 Write a fraction in lowest terms using common factors.

Write each fraction in lowest terms.

13. $\frac{2}{8}$ 14. $\frac{4}{12}$ 15. $\frac{14}{49}$ 16. $\frac{8}{36}$

17. $\frac{25}{30}$ 18. $\frac{10}{18}$ 19. $\frac{10}{35}$ 20. $\frac{16}{56}$

Objective 3 Write a fraction in lowest terms using prime factors.

Write each fraction in lowest terms using prime factors.

21. $\frac{63}{84}$ 22. $\frac{28}{56}$ 23. $\frac{180}{210}$ 24. $\frac{72}{90}$

25. $\frac{36}{54}$ 26. $\frac{71}{142}$ 27. $\frac{75}{500}$ 28. $\frac{96}{132}$

Objective 4 Tell whether two fractions are equivalent.

Decide whether each pair of fractions is **equivalent** *or* **not equivalent**.

29. $\frac{4}{12}$ and $\frac{1}{4}$ 30. $\frac{3}{7}$ and $\frac{6}{14}$ 31. $\frac{2}{3}$ and $\frac{10}{15}$ 32. $\frac{6}{21}$ and $\frac{3}{7}$

33. $\dfrac{8}{16}$ and $\dfrac{15}{20}$ 34. $\dfrac{4}{14}$ and $\dfrac{1}{3}$ 35. $\dfrac{9}{12}$ and $\dfrac{6}{8}$ 36. $\dfrac{8}{36}$ and $\dfrac{2}{9}$

37. $\dfrac{12}{28}$ and $\dfrac{18}{42}$ 38. $\dfrac{20}{24}$ and $\dfrac{15}{31}$ 39. $\dfrac{3}{39}$ and $\dfrac{1}{3}$ 40. $\dfrac{6}{12}$ and $\dfrac{8}{16}$

2.4 Mixed Exercises

Write whether or not each fraction is written in lowest terms.

41. $\dfrac{6}{13}$ 42. $\dfrac{39}{78}$ 43. $\dfrac{46}{49}$ 44. $\dfrac{10}{27}$

Write each fraction in lowest terms.

45. $\dfrac{12}{88}$ 46. $\dfrac{30}{42}$ 47. $\dfrac{28}{98}$ 48. $\dfrac{26}{39}$

Write each fraction in lowest terms using prime factors.

49. $\dfrac{48}{120}$ 50. $\dfrac{90}{108}$ 51. $\dfrac{105}{252}$ 52. $\dfrac{230}{450}$

*Decide whether each pair of fractions is **equivalent** or **not equivalent**.*

53. $\dfrac{24}{72}$ and $\dfrac{30}{90}$ 54. $\dfrac{7}{5}$ and $\dfrac{21}{18}$ 55. $\dfrac{15}{10}$ and $\dfrac{24}{16}$ 56. $\dfrac{5}{35}$ and $\dfrac{10}{60}$

57. $\dfrac{7}{11}$ and $\dfrac{9}{12}$ 58. $\dfrac{12}{28}$ and $\dfrac{15}{35}$ 59. $\dfrac{9}{72}$ and $\dfrac{8}{56}$ 60. $\dfrac{6}{27}$ and $\dfrac{4}{18}$

MULTIPLYING AND DIVIDING FRACTIONS

2.5 Multiplying Fractions

Objective 1   Multiplying fractions.

Multiply. Write the answer in lowest terms.

1. $\dfrac{3}{8} \cdot \dfrac{5}{9}$

2. $\dfrac{5}{9} \cdot \dfrac{7}{6}$

3. $\dfrac{1}{3} \cdot \dfrac{2}{5}$

4. $\dfrac{4}{7} \cdot \dfrac{3}{5}$

5. $\dfrac{5}{6} \cdot \dfrac{11}{4}$

6. $\dfrac{9}{10} \cdot \dfrac{3}{2}$

7. $\dfrac{7}{8} \cdot \dfrac{1}{5}$

8. $\dfrac{10}{7} \cdot \dfrac{4}{5}$

9. $\dfrac{1}{2} \cdot \dfrac{4}{3}$

10. $\dfrac{2}{5} \cdot \dfrac{15}{16}$

11. $\dfrac{1}{9} \cdot \dfrac{2}{3} \cdot \dfrac{5}{6}$

12. $\dfrac{3}{8} \cdot \dfrac{1}{4} \cdot \dfrac{1}{9}$

Objective 2   Use a multiplication shortcut.

Use the multiplication shortcut to find each product. Write the answer in lowest terms.

13. $\dfrac{1}{6} \cdot \dfrac{9}{8}$

14. $\dfrac{7}{6} \cdot \dfrac{3}{14}$

15. $\dfrac{4}{9} \cdot \dfrac{15}{16}$

16. $\dfrac{3}{5} \cdot \dfrac{25}{27}$

17. $\dfrac{11}{4} \cdot \dfrac{8}{33}$

18. $\dfrac{9}{10} \cdot \dfrac{2}{3}$

19. $\dfrac{5}{6} \cdot \dfrac{4}{35}$

20. $\dfrac{10}{7} \cdot \dfrac{63}{21}$

21. $\dfrac{4}{13} \cdot \dfrac{52}{64}$

22. $\dfrac{8}{11} \cdot \dfrac{55}{72}$

23. $\dfrac{3}{4} \cdot \dfrac{5}{9} \cdot \dfrac{2}{5}$

24. $\dfrac{3}{8} \cdot \dfrac{4}{9} \cdot \dfrac{15}{6}$

Objective 3   Multiply a fraction and a whole number.

Multiply. Write the answer in lowest terms; change the answer to a whole or mixed number where possible.

25. $27 \cdot \dfrac{5}{9}$

26. $49 \cdot \dfrac{6}{7}$

27. $72 \cdot \dfrac{5}{8}$

28. $26 \cdot \dfrac{2}{13}$

29. $30 \cdot \dfrac{3}{5}$

30. $27 \cdot \dfrac{7}{54}$

31. $21 \cdot \dfrac{3}{7} \cdot \dfrac{7}{9}$

32. $8 \cdot \dfrac{7}{32} \cdot \dfrac{1}{2}$

33. $200 \cdot \dfrac{7}{50} \cdot \dfrac{5}{28}$

Objective 4   **Find the area of a rectangle.**

Find the area of each rectangle.

34. Length: $\frac{1}{2}$ foot, width: $\frac{1}{3}$ foot

35. Length: $\frac{2}{3}$ yard, width: $\frac{1}{2}$ yard

36. Length: $\frac{3}{4}$ meter, width: $\frac{1}{2}$ meter

37. Length: $\frac{3}{8}$ inch, width: $\frac{4}{11}$ inch

38. Length: $\frac{5}{3}$ yards, width: $\frac{3}{2}$ yards

39. Length: $\frac{8}{11}$ foot, width: $\frac{2}{9}$ foot

40. Length: $\frac{7}{16}$ inch, width: $\frac{3}{16}$ inch

41. Length: $\frac{9}{16}$ meter, width: $\frac{5}{6}$ meter

Solve each application problem.

42. A desk is $\frac{2}{3}$ yard by $\frac{5}{6}$ yard. Find its area.

43. A wading pool is $\frac{5}{4}$ yards by $\frac{5}{9}$ yard. Find its area.

2.5 Mixed Exercises

Multiply. Write the answer in lowest terms.

44. $\dfrac{3}{4} \cdot \dfrac{5}{6} \cdot \dfrac{2}{3}$

45. $\dfrac{3}{4} \cdot \dfrac{8}{9} \cdot \dfrac{7}{10}$

46. $\dfrac{3}{5} \cdot \dfrac{15}{8} \cdot \dfrac{4}{9}$

47. $\dfrac{2}{7} \cdot \dfrac{3}{5} \cdot \dfrac{3}{4}$

Use the multiplication shortcut to find each product. Write the answer in lowest terms.

48. $\dfrac{7}{11} \cdot \dfrac{22}{30} \cdot \dfrac{40}{21}$

49. $\dfrac{15}{16} \cdot \dfrac{6}{5} \cdot \dfrac{2}{3}$

50. $\dfrac{3}{15} \cdot \dfrac{12}{10} \cdot \dfrac{25}{36}$

51. $\dfrac{25}{35} \cdot \dfrac{14}{30} \cdot \dfrac{3}{7}$

Multiply. Write the answer in lowest terms; change the answer to a whole or mixed number where possible.

52. $39 \cdot \dfrac{7}{13} \cdot \dfrac{6}{42}$

53. $6 \cdot \dfrac{7}{300}$

54. $\dfrac{4}{250} \cdot 50$

55. $\dfrac{21}{520} \cdot 13 \cdot \dfrac{20}{7}$

56. $\dfrac{15}{64} \cdot 32 \cdot \dfrac{8}{35}$

57. $\dfrac{48}{39} \cdot 13 \cdot \dfrac{7}{24}$

58. $\dfrac{9}{26} \cdot \dfrac{39}{18} \cdot 12$

Find the area of each rectangle.

59. Length: $\frac{3}{13}$ yard, width: $\frac{2}{7}$ yard

60. Length: $\frac{15}{16}$ foot, width: $\frac{2}{3}$ foot

61. Length: $\frac{11}{8}$ inches, width: $\frac{2}{11}$ inches

MULTIPLYING AND DIVIDING FRACTIONS

2.6 Applications of Multiplication

Objective 1   **Solve application problems using multiplication.**

Solve each application problem. Look for indicator words.

1. A bookstore sold 2800 books, $\frac{3}{5}$ of which were paperbacks. How many paperbacks were sold?

2. A store sells 3750 items, of which $\frac{2}{15}$ are classified as junk food. How many of the items are junk food?

3. Sara needs $2500 to go to school for one year. She earns $\frac{3}{5}$ of this amount in the summer. How much does she earn in the summer?

4. Lany paid $120 for textbooks this term. Of this amount, the bookstore kept $\frac{1}{4}$. How much did the bookstore keep?

5. Of the 570 employees of Grand Tire Service, $\frac{7}{30}$ have given to the United Fund. How many have given to the United Fund?

6. A school gives scholarships to $\frac{3}{25}$ of its 1900 freshmen. How many students receive scholarships?

7. Kim earns $1500 a month. If she uses $\frac{1}{3}$ of her income on housing, how much does she pay for housing?

8. Steve's home is $\frac{3}{5}$ of the way from Carolyn's home to Laplace College, a distance of 45 miles. How far is it from Carolyn's home to Steve's?

The Hu family earned $36,000 last year. Use this fact to solve Problems 9 – 11.

9. They paid $\frac{1}{3}$ of their income for taxes. Find their tax amount.

10. They spend $\frac{2}{5}$ of their income for rent. Find the amount spent on rent.

11. They saved $\frac{1}{16}$ of their income. How much did they save?

MULTIPLYING AND DIVIDING FRACTIONS

2.7 Dividing Fractions

Objective 1 Find the reciprocal of a fraction.

1. $\dfrac{3}{4}$

2. $\dfrac{9}{2}$

3. $\dfrac{1}{3}$

4. $\dfrac{6}{7}$

5. 10

6. $\dfrac{15}{4}$

Objective 2 Divide fractions.

Divide. Write the answer in lowest terms; change the answers to a whole or mixed number where possible.

7. $\dfrac{1}{9} \div \dfrac{1}{3}$

8. $\dfrac{4}{5} \div \dfrac{3}{8}$

9. $\dfrac{\frac{7}{10}}{\frac{14}{5}}$

10. $\dfrac{\frac{4}{9}}{\frac{16}{27}}$

11. $\dfrac{\frac{8}{15}}{\frac{10}{12}}$

12. $\dfrac{28}{5} \div \dfrac{42}{25}$

13. $9 \div \dfrac{3}{2}$

14. $15 \div \dfrac{2}{5}$

15. $\dfrac{5}{8} \div 15$

16. $\dfrac{\frac{6}{11}}{18}$

17. $\dfrac{\frac{11}{3}}{5}$

18. $4 \div \dfrac{12}{7}$

Objective 3 Solve application problems in which fractions are divided.

Solve each application problem by using division.

19. Abel has a piece of property with an area of $\frac{7}{8}$ acre. He wishes to divide it into four equal parts for his children. How many acres of land will each child get?

20. Amanda wants to make doll dresses to sell at a craft's fair. Each dress needs $\frac{1}{3}$ yard of material. She has 18 yards of material. Find the number of dresses that she can make.

21. It takes $\frac{4}{5}$ pound of salt to fill a large salt shaker. How many salt shakers can be filled with 32 pounds of salt?

22. Lynn has 2 gallons of lemonade. If each of her Brownies gets $\frac{1}{12}$ gallon of lemonade, how many Brownies does she have?

23. How many $\frac{1}{9}$-ounce medicine vials can be filled with 7 ounces of medicine?

24. Each guest at a party will eat $\frac{5}{16}$ pound of chips. How many guests can be served with 10 pounds of chips?

25. Samantha uses $\frac{2}{3}$ yard of ribbon to make a bow for each package she wraps at May's Department Store. How many bows can she make if she has 60 yards of ribbon?

2.7 Mixed Exercises

Find the reciprocal of each fraction.

26. $\dfrac{8}{5}$ 27. $\dfrac{4}{11}$ 28. 12

Divide. Write the answer in lowest terms; change the answer to a whole or mixed number where possible.

29. $5 \div \dfrac{13}{20}$ 30. $\dfrac{36}{\frac{12}{25}}$ 31. $\dfrac{11}{\frac{7}{8}}$ 32. $\dfrac{9}{\frac{15}{34}}$

Solve each application problem by using division.

33. Bill wishes to make hamburger patties that weigh $\frac{5}{12}$ pound. How many hamburger patties can he make with 10 pounds of hamburger?

34. Glen has a small pickup truck that will carry $\frac{3}{4}$ cord of firewood. Find the number of trips needed to deliver 30 cords of wood.

35. How many $\frac{5}{4}$-cup glass tumblers can be filled from a 20-cup bowl of punch?

MULTIPLYING AND DIVIDING FRACTIONS

2.8 Multiplying and Dividing Mixed Numbers

Objective 1 Estimate the answer and multiply mixed numbers.

First estimate the answer. Then multiply to find the exact answer. Write the answer as a mixed number or a whole number.

1. $5\frac{1}{3} \cdot 2\frac{1}{2}$

2. $3\frac{1}{2} \cdot 4\frac{2}{7}$

3. $5\frac{1}{4} \cdot 3\frac{1}{5}$

4. $3\frac{1}{2} \cdot 1\frac{3}{7}$

5. $4\frac{4}{9} \cdot 2\frac{2}{5}$

6. $5\frac{2}{3} \cdot 7\frac{1}{8}$

7. $2\frac{1}{6} \cdot 3\frac{3}{4}$

8. $18 \cdot 2\frac{5}{9}$

9. $6 \cdot 3\frac{1}{2}$

10. $3\frac{2}{5} \cdot 15$

11. $\frac{5}{6} \cdot 2\frac{1}{2} \cdot 2\frac{2}{5}$

12. $1\frac{1}{4} \cdot 1\frac{1}{3} \cdot 1\frac{1}{2}$

Objective 2 Estimate the answer and divide mixed numbers.

First estimate the answer. Then divide to find the exact answer. Write the answer as a mixed number or a whole number.

13. $5\frac{5}{6} \div 5\frac{1}{4}$

14. $3\frac{1}{8} \div 2\frac{3}{4}$

15. $4\frac{5}{8} \div 1\frac{1}{4}$

16. $4\frac{3}{8} \div 3\frac{1}{2}$

17. $5\frac{3}{5} \div 1\frac{1}{6}$

18. $6\frac{1}{4} \div 2\frac{1}{2}$

19. $5 \div 3\frac{3}{4}$

20. $14 \div 8\frac{2}{5}$

21. $7\frac{1}{3} \div 6$

22. $4\frac{2}{3} \div 2$

23. $7\frac{1}{2} \div \frac{2}{3}$

24. $2\frac{5}{8} \div 1\frac{3}{4}$

Objective 3 Solve application problems with mixed numbers.

First estimate the answer. Then solve each application problem by using multiplication or division to find the exact answer.

25. Maria wants to make 20 dresses to sell at a bazaar. Each dress needs $3\frac{1}{4}$ yards of material. How many yards does she need?

26. Juan worked $38\frac{1}{4}$ hours at $9 per hour. How much did he make?

27. Each home in an area needs $41\frac{1}{3}$ yards of rain gutter. How much rain gutter would be needed for 6 homes?

28. A farmer applies fertilizer to this fields at a rate of $5\frac{5}{6}$ gallons per acre. How many acres can he fertilize with $65\frac{5}{6}$ gallons?

29. Insect spray is mixed using $1\frac{3}{4}$ ounces of a chemical per gallon of water. How many ounces of the chemical are needed to mix with $28\frac{4}{5}$ gallons of water?

30. How many $\frac{3}{4}$-pound peanut cans can be filled with 15 pounds of peanuts?

2.8 Mixed Exercises

First estimate the answer. Then multiply or divide to find the exact answer. Write the answer as a mixed number or a whole number.

31. $9 \cdot 3\frac{1}{4} \cdot 1\frac{3}{13} \cdot 2\frac{2}{3}$ 32. $3\frac{1}{4} \cdot 3\frac{3}{13} \cdot 1\frac{5}{12}$ 33. $4\frac{2}{7} \cdot 2\frac{2}{9} \cdot 2\frac{4}{5}$ 34. $3\frac{1}{8} \cdot 4\frac{4}{5} \cdot 5\frac{2}{3}$

35. $3\frac{1}{6} \div 1\frac{2}{3}$ 36. $8\frac{3}{4} \div 5$ 37. $16 \div 2\frac{7}{8}$ 38. $6\frac{3}{8} \div 4\frac{1}{4}$

First estimate the answer. Then solve each application problem by using multiplication or division to find the exact answer.

39. How many dresses can be made from 70 yards of material if each dress requires $4\frac{3}{8}$ yards?

40. Arnette worked $24\frac{1}{2}$ hours and earned \$9 in one hour. How much did she make?

41. Juan has $3\frac{1}{2}$ sticks of margarine. If each stick weighs $\frac{1}{4}$ pound, how much does Juan's margarine weigh?

42. A dental office plays taped music constantly. Each tape takes $1\frac{1}{4}$ hours. How many tapes are played during $7\frac{1}{2}$ hours?

Chapter 3

ADDING AND SUBTRACTING FRACTIONS

3.1 Adding and Subtracting Like Fractions

Objective 1   Define like and unlike fractions.

Write **like** *or* **unlike** *for each set of fractions.*

1. $\dfrac{9}{7}, \dfrac{2}{7}$

2. $\dfrac{11}{9}, \dfrac{4}{9}$

3. $\dfrac{3}{5}, \dfrac{4}{10}$

4. $\dfrac{4}{9}, \dfrac{4}{7}$

5. $\dfrac{2}{5}, \dfrac{3}{5}$

6. $\dfrac{1}{6}, \dfrac{5}{6}, \dfrac{7}{12}$

7. $\dfrac{2}{3}, \dfrac{3}{2}$

8. $\dfrac{9}{9}, \dfrac{4}{9}, \dfrac{3}{8}$

9. $\dfrac{2}{15}, \dfrac{3}{15}, \dfrac{1}{5}$

10. $\dfrac{4}{11}, \dfrac{11}{11}, \dfrac{22}{11}$

11. $\dfrac{0}{3}, \dfrac{5}{3}, \dfrac{1}{3}$

12. $\dfrac{18}{7}, \dfrac{21}{7}, \dfrac{7}{7}$

Objective 2   Add like fractions.

Add. Write the answer in lowest terms and as a mixed number when possible.

13. $\dfrac{1}{4} + \dfrac{3}{4}$

14. $\dfrac{3}{7} + \dfrac{2}{7}$

15. $\dfrac{5}{8} + \dfrac{1}{8}$

16. $\dfrac{11}{15} + \dfrac{1}{15}$

17. $\dfrac{7}{8} + \dfrac{5}{8}$

18. $\dfrac{4}{3} + \dfrac{7}{3}$

19. $\dfrac{11}{16} + \dfrac{7}{16}$

20. $\dfrac{9}{8} + \dfrac{1}{8}$

21. $\dfrac{1}{6} + \dfrac{5}{6}$

22. $\dfrac{1}{5} + \dfrac{2}{5} + \dfrac{4}{5}$

23. $\dfrac{6}{10} + \dfrac{4}{10} + \dfrac{3}{10}$

24. $\dfrac{3}{49} + \dfrac{7}{49} + \dfrac{18}{49}$

Solve each application problem. Write the answer in lowest terms.

25. Last month the Yee family paid $\frac{2}{11}$ of a debt. This month they paid an additional $\frac{5}{11}$ of the same debt. What fraction of the debt has been paid?

26. Malika walked $\frac{3}{8}$ of a mile downhill and then $\frac{1}{8}$ of a mile along a creek. How far did she walk altogether?

Objective 3   Subtract like fractions.

Subtract. Write the answer in lowest terms and as a mixed number when possible.

27. $\dfrac{11}{13} - \dfrac{3}{13}$ 28. $\dfrac{3}{10} - \dfrac{1}{10}$ 29. $\dfrac{15}{7} - \dfrac{8}{7}$ 30. $\dfrac{11}{16} - \dfrac{3}{16}$

31. $\dfrac{7}{8} - \dfrac{3}{8}$ 32. $\dfrac{16}{21} - \dfrac{2}{21}$ 33. $\dfrac{16}{15} - \dfrac{6}{15}$ 34. $\dfrac{8}{25} - \dfrac{5}{25}$

35. $\dfrac{29}{35} - \dfrac{1}{35}$ 36. $\dfrac{25}{28} - \dfrac{15}{28}$ 37. $\dfrac{31}{36} - \dfrac{11}{36}$ 38. $\dfrac{73}{90} - \dfrac{58}{90}$

Solve each application problem. Write the answer in lowest terms.

39. Bill must walk $\frac{9}{12}$ of a mile. He has already walked $\frac{1}{12}$ of a mile. How much farther must he walk?

40. The Thompsons still owe $\frac{8}{15}$ of a debt. If they pay $\frac{2}{15}$ of it this month, what fraction of the debt will they still owe?

3.1 Mixed Exercises

Write like or unlike for each set of fractions.

41. $\dfrac{1}{2}, \dfrac{2}{4}, \dfrac{4}{2}$ 42. $\dfrac{2}{3}, \dfrac{7}{3}, \dfrac{11}{3}, \dfrac{1}{3}$ 43. $\dfrac{4}{10}, \dfrac{7}{10}, \dfrac{9}{10}, \dfrac{1}{10}$ 44. $\dfrac{9}{4}, \dfrac{5}{4}, \dfrac{4}{3}, \dfrac{8}{4}$

Add or subtract as indicated. Write the answer in lowest terms and as a mixed number when possible.

45. $\dfrac{2}{13} + \dfrac{15}{13} + \dfrac{9}{13}$ 46. $\dfrac{15}{72} + \dfrac{19}{72} + \dfrac{14}{72}$ 47. $\dfrac{67}{81} + \dfrac{29}{81} + \dfrac{12}{81}$

48. $\dfrac{91}{100} - \dfrac{41}{100}$ 49. $\dfrac{736}{400} - \dfrac{496}{400}$ 50. $\dfrac{365}{224} - \dfrac{269}{224}$

Solve each application problem. Write the answer in lowest terms.

51. Brent painted $\frac{1}{6}$ of a house last week and another $\frac{3}{6}$ this week. How much of the house is painted?

52. Jeff planted $\frac{11}{18}$ of his garden in corn and potatoes. If $\frac{5}{18}$ of the garden is corn, how much of the garden is potatoes?

ADDING AND SUBTRACTING FRACTIONS

3.2 Least Common Multiples

Objective 1   **Find the least common multiple.**

Find the least common multiple for each of the following by listing the common multiples of each number.

1.   7, 14

2.   8, 16

3.   12, 18

4.   14, 21

5.   21, 28

6.   20, 35

7.   20, 65

8.   40, 50

9.   15, 21

Objective 2   **Find the least common multiple by using multiples of the largest number.**

Find the least common multiple for each of the following by using multiples of the larger number.

10. 5, 12

11. 16, 20

12. 8, 12

13. 15, 25

14. 15, 18

15. 14, 35

16. 6, 9

17. 21, 28

18. 32, 40

Objective 3   **Find the least common multiple by using prime factorization.**

Find the least common multiple for each of the following using prime factorization.

19. 14, 28

20. 16, 18

21. 28, 32

22. 25, 35

23. 10, 24, 32

24. 15, 18, 20

25. 16, 20, 25

26. 7, 12, 21, 35

27. 8, 14, 24, 40

Objective 4   **Find the least common multiple by using an alternate method.**

Find the least common multiple for each set of numbers by using an alternate method.

28. 3, 12

29. 9, 18

30. 10, 15

31. 22, 55

32. 32, 36

33. 26, 65

34. 35, 85

35. 4, 18, 27

36. 8, 24, 36

37. 12, 30, 40

38. 7, 20, 35

39. 10, 12, 36

Objective 5   **Write a fraction with an indicted denominator.**

Rewrite each fraction with the indicated denominator.

40. $\dfrac{1}{3} = \dfrac{}{12}$

41. $\dfrac{1}{2} = \dfrac{}{28}$

42. $\dfrac{1}{9} = \dfrac{}{36}$

43. $\dfrac{2}{7} = \dfrac{}{63}$

44. $\dfrac{4}{7} = \dfrac{}{35}$

45. $\dfrac{4}{11} - \dfrac{}{132}$

46. $\dfrac{4}{9} = \dfrac{}{81}$

47. $\dfrac{6}{5} = \dfrac{}{75}$

48. $\dfrac{1}{13} = \dfrac{}{78}$

49. $\dfrac{5}{6} = \dfrac{}{72}$

50. $\dfrac{3}{8} = \dfrac{}{88}$

51. $\dfrac{3}{13} = \dfrac{}{52}$

52. $\dfrac{7}{12} = \dfrac{}{60}$

53. $\dfrac{7}{20} = \dfrac{}{80}$

54. $\dfrac{21}{11} = \dfrac{}{55}$

3.2 Mixed Exercises

Find the least common multiple for each of the following by listing the common multiples of each number.

55. 5, 15, 25   56. 7, 8, 56   57. 6, 9, 45   58. 3, 12, 15

Find the least common multiple for each of the following by using multiples of the larger number.

59. 12, 20   60. 18, 24   61. 9, 16

Find the least common multiple for each of the following using prime factorization.

62. 9, 12, 36, 45   63. 10, 25, 30, 52   64. 6, 27, 42, 63, 84

Find the least common multiple for each set of numbers by using an alternative method.

65. 15, 20, 35, 48   66. 9, 15, 25, 27   67. 20, 36, 42, 50

Rewrite each fraction with the indicated denominator.

68. $\dfrac{15}{7} = \dfrac{}{84}$

69. $\dfrac{18}{11} = \dfrac{}{77}$

70. $\dfrac{19}{21} = \dfrac{}{105}$

71. $\dfrac{12}{15} = \dfrac{}{135}$

72. $\dfrac{9}{17} = \dfrac{}{102}$

ADDING AND SUBTRACTING FRACTIONS

3.3 Adding and Subtracting Unlike Fractions

Objective 1 Add unlike fractions.

Add. Write the answer in lowest terms.

1. $\dfrac{2}{3}+\dfrac{1}{6}$ 2. $\dfrac{1}{3}+\dfrac{1}{2}$ 3. $\dfrac{1}{5}+\dfrac{5}{8}$ 4. $\dfrac{9}{13}+\dfrac{3}{26}$

5. $\dfrac{3}{10}+\dfrac{7}{15}$ 6. $\dfrac{5}{8}+\dfrac{1}{4}$ 7. $\dfrac{3}{11}+\dfrac{2}{33}$ 8. $\dfrac{5}{12}+\dfrac{9}{16}$

9. $\dfrac{3}{5}+\dfrac{2}{9}$ 10. $\dfrac{3}{8}+\dfrac{5}{12}$

Solve each application problem.

11. A buyer for a grain company bought $\frac{3}{8}$ ton of wheat, $\frac{1}{6}$ ton of rice, and $\frac{1}{4}$ ton of barley. How many tons of grain were bought?

12. Michael Pippen paid $\frac{1}{9}$ of a debt in January, $\frac{1}{2}$ in February, $\frac{1}{4}$ in March, and $\frac{1}{12}$ in April. What fraction of the debt was paid in these four months?

Objective 2 Add fractions vertically.

Add. Write the answer in lowest terms.

13. $\dfrac{1}{2}$ 14. $\dfrac{7}{12}$ 15. $\dfrac{2}{15}$ 16. $\dfrac{1}{6}$

$+\dfrac{1}{3}$ $+\dfrac{3}{8}$ $+\dfrac{7}{10}$ $+\dfrac{2}{9}$

17. $\dfrac{3}{7}$ 18. $\dfrac{5}{22}$ 19. $\dfrac{6}{13}$ 20. $\dfrac{3}{14}$

$+\dfrac{4}{21}$ $+\dfrac{7}{33}$ $+\dfrac{15}{52}$ $+\dfrac{5}{21}$

Objective 3   **Subtract unlike fractions.**

Subtract. Write the answer in lowest terms.

21. $\dfrac{7}{8}-\dfrac{1}{2}$ 22. $\dfrac{11}{12}-\dfrac{7}{18}$ 23. $\dfrac{2}{3}-\dfrac{1}{6}$ 24. $\dfrac{5}{8}-\dfrac{1}{6}$ 25. $\dfrac{7}{12}-\dfrac{1}{4}$

26. $\dfrac{8}{15}-\dfrac{1}{5}$

27. $\begin{array}{r}\dfrac{7}{8}\\[4pt]-\dfrac{2}{3}\\\hline\end{array}$

28. $\begin{array}{r}\dfrac{5}{9}\\[4pt]-\dfrac{5}{12}\\\hline\end{array}$

29. $\begin{array}{r}\dfrac{7}{8}\\[4pt]-\dfrac{5}{6}\\\hline\end{array}$

30. $\begin{array}{r}\dfrac{2}{3}\\[4pt]-\dfrac{3}{5}\\\hline\end{array}$

31. $\begin{array}{r}\dfrac{7}{15}\\[4pt]-\dfrac{3}{10}\\\hline\end{array}$

32. $\begin{array}{r}\dfrac{5}{12}\\[4pt]-\dfrac{1}{4}\\\hline\end{array}$

Solve each application problem.

33. A company has $\frac{5}{8}$ acre of land. They sell $\frac{1}{3}$ acre. How much land is left?

34. Greg had $\frac{7}{12}$ of his savings goal to complete at the beginning of the month. During the month he saved another $\frac{1}{8}$ of the goal. How much of the goal is left to save?

3.3 Mixed Exercises

Add or subtract as indicated. Write the answer in lowest terms.

35. $\dfrac{1}{6}+\dfrac{3}{14}$ 36. $\dfrac{4}{15}+\dfrac{9}{20}$ 37. $\dfrac{1}{4}+\dfrac{2}{7}+\dfrac{3}{14}$ 38. $\dfrac{1}{3}+\dfrac{1}{8}+\dfrac{5}{12}$

39. $\begin{array}{r}\dfrac{5}{18}\\[4pt]+\dfrac{7}{27}\\\hline\end{array}$

40. $\begin{array}{r}\dfrac{1}{12}\\[4pt]+\dfrac{1}{8}\\\hline\end{array}$

41. $\dfrac{8}{9}-\dfrac{2}{15}$ 42. $\dfrac{5}{12}-\dfrac{7}{18}$

43. $\begin{array}{r}\dfrac{9}{16}\\[4pt]-\dfrac{3}{10}\\\hline\end{array}$

44. $\begin{array}{r}\dfrac{7}{8}\\[4pt]-\dfrac{7}{28}\\\hline\end{array}$

45. $\dfrac{3}{5}-\dfrac{1}{4}$ 46. $\dfrac{9}{10}-\dfrac{4}{25}$

ADDING AND SUBTRACTING FRACTIONS

3.4 Adding and Subtracting Mixed Numbers

Objective 1 Estimate an answer, then add or subtract mixed numbers.

First estimate the answer. Then add or subtract to find the exact answer. Write each answer as a mixed number in lowest terms.

1. $5\frac{1}{7}$
    $+\ 4\frac{3}{7}$

2. $3\frac{1}{9}$
    $+\ 4\frac{7}{8}$

3. $7\frac{3}{4}$
    $+\ 4\frac{5}{8}$

4. $17\frac{5}{8}$
    $12\frac{1}{4}$
    $+\ \ 5\frac{5}{6}$

5. $59\frac{7}{8}$
    $24\frac{5}{6}$
    $+\ 13\frac{1}{12}$

6. $126\frac{4}{5}$
    $28\frac{9}{10}$
    $+\ 13\frac{2}{15}$

7. $5\frac{3}{5}$
    $-\ 2\frac{1}{10}$

8. $12\frac{9}{16}$
    $-\ 2\frac{3}{8}$

9. $9\frac{7}{12}$
    $-\ 2\frac{1}{3}$

First estimate the answer. Then solve each application problem.

10. A painter used $2\frac{1}{3}$ cans of paint one day and $1\frac{7}{8}$ cans the next day. How many cans did he use altogether?

11. A mechanic had $8\frac{3}{4}$ gallons of transmission fluid. If he purchased $2\frac{1}{3}$ gallons of fluid, find the number of gallons he has altogether.

12.   Nate bought $2\frac{3}{8}$ boxes of oranges and $2\frac{2}{3}$ boxes of lemons. How many boxes of fruit did he buy in all?

13. On Monday, $7\frac{3}{4}$ tons of cans were recycled, while $9\frac{3}{5}$ tons were recycled on Tuesday. How many tons were recycled in total on these two days?

14. Paul worked $12\frac{3}{4}$ hours over the weekend. He worked $6\frac{3}{8}$ hours on Saturday. How many hours did he work on Sunday?

15. Marty Hirsch worked $6\frac{2}{5}$ hours on Monday, $7\frac{1}{2}$ hours on Tuesday, $8\frac{3}{4}$ hours on Wednesday, $7\frac{4}{5}$ hours on Thursday, and 8 hours on Friday. How many hours did he work altogether?

16. The Eastside Wholesale Vegetable Market sold $4\frac{3}{4}$ tons of broccoli, $8\frac{2}{3}$ tons of spinach, $2\frac{1}{2}$ tons of corn, and $1\frac{5}{12}$ tons of turnips last month. Find the total number of tons of these vegetables sold by the market last month.

| Objective 2 | **Estimate an answer, then subtract mixed numbers by borrowing.**

First estimate the answer. Then subtract to find the exact answer. Write each answer as a mixed number in lowest terms.

17. $9\frac{1}{8}$
     $-7\frac{3}{8}$

18. $11\frac{1}{4}$
     $-\ 6\frac{3}{4}$

19. $6\frac{1}{3}$
     $-5\frac{7}{12}$

20. $7\frac{9}{20}$
     $-5\ \frac{3}{5}$

21. $27\frac{2}{15}$
     $-18\frac{7}{10}$

22. $12\frac{5}{12}$
     $-11\frac{11}{16}$

23. $42\frac{7}{12}$
     $-29\frac{5}{8}$

24. 42
     $-19\frac{3}{4}$

25. 21
     $-17\frac{9}{16}$

First estimate the answer. Then solve each application problem.

26. Amy Atwood worked 40 hours during a certain week. She worked $8\frac{1}{4}$ hours on Monday, $6\frac{3}{8}$ hours on Tuesday, $7\frac{3}{4}$ hours on Wednesday, and $8\frac{3}{4}$ hours on Thursday. How many hours did she work on Friday?

27. Three sides of a parking lot are $35\frac{1}{4}$ yards, $42\frac{7}{8}$ yards, and $32\frac{3}{4}$ yards. If the total distance around the lot is $145\frac{1}{2}$ yards, find the length of the fourth side.

28. A concrete truck is loaded with $11\frac{5}{8}$ cubic yards of concrete. The driver unloads $1\frac{1}{6}$ cubic yards at the first stop, and $2\frac{5}{12}$ cubic yards at the second stop. The customer at the third stop gets 3 cubic yards. How much concrete is left in the truck?

29. Debbie Andersen bought 15 yards of material at a sale. She made a shirt with $3\frac{1}{8}$ yards of the material, a dress with $4\frac{7}{8}$ yards, and a jacket with $3\frac{3}{4}$ yards. How many yards of material were left over?

Find x in the following figures.

30.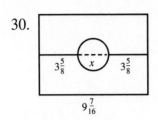

31.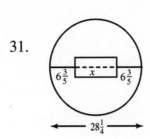

Add or subtract mixed numbers using an alternative method.

Add or subtract by changing mixed numbers to improper fractions. Write the answer as a mixed number.

32. $3\frac{3}{4}$
     $+1\frac{1}{2}$

33. $5\frac{1}{3}$
     $+2\frac{5}{6}$

34. $1\frac{3}{8}$
     $+2\frac{3}{5}$

35. $3\frac{7}{8}$
     $+1\frac{5}{12}$

36. $4\frac{3}{4}$
     $-2\frac{3}{8}$

37. $3\frac{1}{2}$
     $-1\frac{2}{3}$

38. $5\frac{5}{8}$
     $-2\frac{3}{4}$

39. $3\frac{2}{3}$
     $-1\frac{5}{6}$

40. $4\frac{1}{8}$
     $-2\frac{5}{12}$

3.3 Mixed Exercises

First estimate the answer. Then add or subtract as indicated. Write each answer as a mixed number in lowest terms.

41. $22\frac{3}{5}$
     $+15\frac{4}{5}$

42. $14\frac{5}{6}$
     $+8\frac{5}{6}$

43. $9\frac{7}{8}$
     $+8\frac{1}{2}$

44. $27\frac{1}{4}$
     $15\frac{3}{8}$
     $+9\frac{1}{2}$

45. $28\frac{3}{5}$
     $47\frac{7}{10}$
     $+23\frac{8}{15}$

46. $28\frac{3}{4}$
     $21\frac{1}{5}$
     $+19\frac{9}{10}$

47. $7\frac{5}{8}$
     $-6\frac{7}{12}$

48. $15\frac{15}{16}$
     $-8\frac{13}{24}$

49. $26\frac{11}{14}$
     $-13\frac{5}{18}$

50. $372\frac{5}{6}$
     $-208\frac{3}{8}$

51. $9\frac{1}{8}$
     $-7\frac{4}{9}$

52. $14\frac{4}{7}$
     $-8\frac{1}{8}$

53. $14\frac{4}{5}$
     $-12\frac{4}{15}$

54. 29
     $-8\frac{1}{12}$

55. $129\frac{2}{3}$
     $-98\frac{14}{15}$

56. 147
     $-39\frac{5}{6}$

57.   28
      $- 4\frac{3}{7}$

58.   75
      $- \frac{5}{8}$

ADDING AND SUBTRACTING FRACTIONS

3.5 Order Relations and the Order of Operations

Objective 1 Identify the greater of two fractions.

Write < or > to make a true statement.

1. $\dfrac{1}{2}$ —— $\dfrac{5}{8}$

2. $\dfrac{2}{3}$ —— $\dfrac{5}{6}$

3. $\dfrac{3}{8}$ —— $\dfrac{5}{16}$

4. $\dfrac{7}{5}$ —— $\dfrac{19}{15}$

5. $\dfrac{5}{12}$ —— $\dfrac{3}{5}$

6. $\dfrac{11}{15}$ —— $\dfrac{13}{20}$

7. $\dfrac{13}{24}$ —— $\dfrac{23}{36}$

8. $\dfrac{23}{40}$ —— $\dfrac{17}{30}$

9. $\dfrac{17}{25}$ —— $\dfrac{9}{16}$

Objective 2   Use exponents with fractions.

Evaluate each of the following.

10. $\left(\dfrac{1}{4}\right)^3$

11. $\left(\dfrac{1}{2}\right)^2$

12. $\left(\dfrac{2}{3}\right)^2$

13. $\left(\dfrac{1}{5}\right)^2$

14. $\left(\dfrac{5}{3}\right)^3$

15. $\left(\dfrac{1}{5}\right)^3$

16. $\left(\dfrac{1}{2}\right)^4$

17. $\left(\dfrac{3}{2}\right)^4$

18. $\left(\dfrac{3}{7}\right)^3$

19. $\left(\dfrac{3}{4}\right)^3$

20. $\left(\dfrac{8}{11}\right)^2$

21. $\left(\dfrac{8}{15}\right)^2$

22. $\left(\dfrac{2}{9}\right)^3$

23. $\left(\dfrac{12}{7}\right)^2$

24. $\left(\dfrac{4}{3}\right)^5$

Objective 3   Use the order of operations.

Use the order of operations to simplify.

25. $\left(\dfrac{2}{3}\right)^2 \cdot 6$

26. $4\cdot\left(\dfrac{1}{8}\right)^2$

27. $\left(\dfrac{4}{5}\right)^2 \cdot \dfrac{5}{12}$

28. $\dfrac{8}{15} \cdot \left(\dfrac{1}{2}\right)^2$

29. $\left(\dfrac{3}{5}\right)^2 \cdot \left(\dfrac{2}{3}\right)^2$

30. $\left(\dfrac{4}{3}\right)^2 \cdot \left(\dfrac{1}{8}\right)^2$

31. $7 \cdot \left(\dfrac{2}{7}\right)^2 \cdot \left(\dfrac{1}{4}\right)^2$

32. $9 \cdot \left(\dfrac{3}{2}\right)^2 \cdot \left(\dfrac{1}{6}\right)^2$

33. $\dfrac{4}{3} - \dfrac{1}{2} + \dfrac{7}{12}$

34. $\dfrac{7}{8} - \dfrac{3}{4} + \dfrac{1}{2}$

35. $\dfrac{1}{3} \cdot \dfrac{3}{7} \cdot \dfrac{5}{4}$

36. $\dfrac{2}{5} \cdot \dfrac{15}{11} \cdot \dfrac{33}{8}$

37. $\dfrac{1}{2} \cdot \dfrac{4}{5} + \dfrac{2}{3} \cdot \dfrac{9}{5}$

38. $\dfrac{3}{14} \cdot \dfrac{7}{5} + \dfrac{1}{2} \cdot \dfrac{2}{5}$

39. $\dfrac{5}{8} - \dfrac{2}{3} \cdot \dfrac{3}{4}$

3.5 Mixed Exercises

Write < or > to make a true statement.

40. $\dfrac{3}{11} \underline{\quad} \dfrac{5}{12}$

41. $\dfrac{7}{9} \underline{\quad} \dfrac{8}{11}$

42. $\dfrac{23}{27} \underline{\quad} \dfrac{51}{59}$

Use the order of operations to simplify.

43. $\dfrac{1}{3} + \left(\dfrac{2}{3}\right)^2 - \dfrac{2}{9}$

44. $\dfrac{2}{9} + \left(\dfrac{2}{9}\right)^2 - \dfrac{5}{27}$

45. $\left(\dfrac{1}{4} + \dfrac{1}{2}\right) \cdot \dfrac{1}{3}$

46. $\dfrac{3}{4} \cdot \left(\dfrac{4}{5} + \dfrac{3}{10}\right)$

47. $\left(\dfrac{8}{7} - \dfrac{9}{14}\right) \div \dfrac{3}{7}$

48. $\dfrac{4}{9} \div \left(\dfrac{5}{9} - \dfrac{1}{3}\right)$

49. $\left(\dfrac{3}{4}\right)^2 + \dfrac{1}{3} \cdot \dfrac{9}{8}$

50. $\left(\dfrac{2}{3}\right)^2 - \dfrac{3}{8} \cdot \dfrac{5}{6}$

51. $\left(\dfrac{3}{5}\right)^2 \cdot \left(\dfrac{1}{3} + \dfrac{2}{9}\right) - \dfrac{1}{5} \cdot \dfrac{5}{8}$

52. $\left(\dfrac{4}{7}\right)^2 \cdot \left(\dfrac{3}{2} - \dfrac{5}{8}\right) - \dfrac{1}{21} \cdot \dfrac{3}{4}$

Chapter 4

DECIMALS

4.1 Reading and Writing Decimals

Objective 1   **Write parts of a whole using decimals.**

Write the portion of each square that is shaded as a fraction, as a decimal, and in words.

1.

2.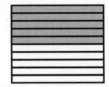

3.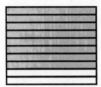

4.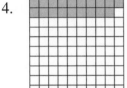

5.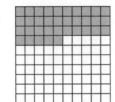

6.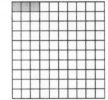

7.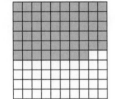

8.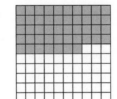

Objective 2   **Identify the place value of a digit.**

Identify the digit that has the given place value.

9. 43.507   tenths   hundredths

10. 0.4952   tenths   hundredths

11. 5.632   tenths   hundredths

12. 0.0478   hundredths   thousandths

13. 0.769   hundredths   thousandths

14. 0.42583   hundredths   thousandths

15. 2.83714   thousandths ten-thousandths

16. 0.78257   thousandths ten-thousandths

17. 42.692   tens   tenths

18. 62.436   tens   tenths

19. 302.9651   hundreds   hundredths

Identify the place value of each digit in these decimals.

20. 0.73   7   3   21. 0.51   5   1   22. 0.85   8   5

23. 0.36   3   6   24. 0.782   7   8   2   25. 0.176   1   7   6

Objective 3   **Read decimals.**

Tell how to read each decimal in words.

26. 0.08   27. 0.007   28. 4.06   29. 3.0014

30. 0.0561   31. 10.835   32. 2.304   33. 97.008

Write each decimal in numbers.

34. Five and four hundredths

35. Eleven and nine thousandths

36. Thirty-eight and fifty-two hundred-thousandths

37. Three hundred and twenty-three ten-thousandths

Objective 4   **Write decimals as fractions or mixed numbers.**

Write each decimal as a fraction or mixed number in lowest terms.

38. 0.8   39. 0.1   40. 3.6   41. 0.5

42. 4.26   43. 0.95   44. 1.66   45. 0.99

46. 3.75

4.1 Mixed Exercises

Write the portion of each square that is shaded as a fraction, as a decimal, and in words.

47.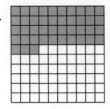

48.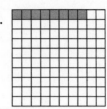

Identify the digit that has the given place value in each decimal.

49. 4569.823   hundreds   hundredths

50. 2019.463   thousands   thousandths

51. 6528.7941   tens   tenths

Identify the place value of each digit given in these decimals.

52. 0.502   5   0   2   53. 0.071   0   7   1

54. 62.51   6   2   5   1   55. 69.435   6   9   4   3   5

Tell how to read each decimal in words.

56   4.083   57. 0.049

Write each decimal in numbers.

58.   Ten and six ten-thousandths   59. Three thousand, six hundred twelve and
                                             seventeen thousandths

Write each decimal as a fraction or a mixed number. Write in lowest terms.

60. 0.89   61. 0.04   62. 4.08   63. 0.504   64. 6.465

DECIMALS

4.2 Rounding Decimals

Objective 2 **Round decimals to any given place.**

Round each number to the place indicated.

1. 17.8937 to the nearest tenth

2. 489.84 to the nearest tenth

3. 785.4982 to the nearest thousandth

4. 43.51499 to the nearest ten-thousandth

5. 53.329 to the nearest hundredth

6. 75.399 to the nearest tenth

Round to the nearest hundredth and then to the nearest tenth. Remember to always round the original number.

7. 89.525

8. 21.769

9. 0.8948

10. 1.437

11. 0.0986

12. 114.038

13. 101.749

14. 78.695

15. 108.073

Objective 3 **Round money amounts to the nearest cent or nearest dollar.**

Round to the nearest dollar.

16. $79.12

17. $28.39

18. $225.98

19. $4797.50

20. $11,839.73

21. $27,869.57

Round to the nearest cent.

22. $1.2499

23. $1.0924

24. $112.0089

25. $134.20506

26. $1028.6666

27. $2096.0149

4.2 Mixed Exercises

Round each number to the place indicated.

28. 38.90952 to the nearest thousandth

29. 10.3228 to the nearest hundredth

30. 799.792 to the nearest tenth

31. 486.496 to the nearest one

Round to the nearest hundredth and then to the nearest tenth. Remember to always round the original number.

32. 3257.595   33. 486.932   34. 264.993   35. 304.857

36. 27.569

Round to the nearest dollar.

37. $55.42   38. $276.49   39. $2460.75

Round to the nearest cent.

40. $62.179   41. $495.6234   42. $1.4948

DECIMALS

4.3 Adding and Subtracting Decimals

Objective 1 Add decimals.

Find each sum.

1. 43.96 + 48.53

2. 47.94 + 102.38 + 27.631

3. 39.87 + 25.2 + 40.36

4. 87.6 + 90.4

5. 45.83 + 20.923 + 5.7

6. 4 + 7.99 + 3.46

7. 10.82 + 5.9 + 4.7 + 6.3 + 20.63

Find the perimeter of (distance around) each geometric figure by adding the lengths of the sides.

8.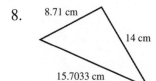

9.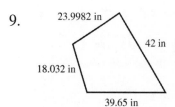

Objective 2   Subtract decimals.

Find each difference.

10. 84.6 – 18.1

11. 223.3 – 107.5

12. 41.2 – 8.76

13. 69.524 – 26.958

14. 23.104 – 6.98

15. 71 – 12.68

16. 689 – 79.832

Find the unknown measurement in each figure.

17.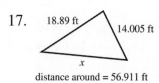

18.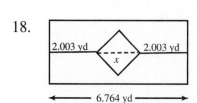

Objective 3   **Estimate the answer when adding or subtracting decimals.**

First, use front end rounding and estimate each answer. Then add or subtract to find the exact answer.

19.   32.99
    41.72
    + 8.2

20.   20.85
    − 7.69

21.   9.7
    − 4.862

22.   593.8
    27.93
    + 54.87

23.   9
    − 3.47

First, use front end rounding and estimate each answer. Then add or subtract to find the exact answer for each application problem.

24. Kim spent $28.25 for books, $29.47 for a blouse, and $17.85 for a compact disk. How much did she spend?

25. Tom Rodriguiz made $365.29 at the regular rate of pay and $87.59 at the overtime rate. How much did he make?

26. Manuel has agreed to work 27.5 hours at a certain job. He has already worked 9.65 hours. How many hours does he have left to work?

27. Michael Lee worked 3.5 days one week, 5.1 days another week, and 4.8 days a third week. How many days did he work altogether?

28. A customer gives a clerk a $20 bill to pay for $11.29 in purchases. How much change should the customer get?

29. A man buys $37.57 worth of sporting goods and pays with a $50 bill. How much change should he get?

30. At a fruit stand, Lynn Knight bought $8.53 worth of apples, $11.10 worth of peaches, and $28.29 worth of pears. How much did she spend altogether?

31. A man receives a bill for $83.26 from Exxon. Of this amount, $53.29 is for a tune-up and the rest is for gas. How much did he pay for gas?

32. At the beginning of a trip to El Cerrito, a car odometer read 80,447.5 miles. It is 81.9 miles to El Cerrito. What should the odometer read after driving to El Cerrito and back?

4.3 Mixed Exercises

Find each sum or difference.

33. $3.09 + 7.54$

34. $49.9 - 38.918$

35. $65 - 0.294$

36. $40.876 + 29.3 + 5.829 + 9.43 + 6.9$

37. $590.87 + 32.53 + 20.4 + 9.6 + 873.1$

38. $\quad\ \ 58.769$
$\underline{-\ 31.8\ \ \ \ }$

39. $\quad\ \ 26.8$
$\underline{-\ 3.963}$

40. $\quad\ \ 51.3$
$\underline{-\ 17.943}$

Find the perimeter of (distance around) each geometric figure by adding the lengths of the sides.

41.

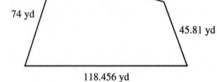

42.

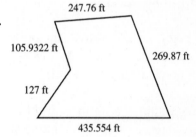

Find the unknown measurement in each figure.

43.

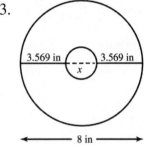

44.

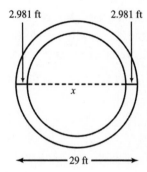

First, use front end rounding and estimate each answer. Then add or subtract to find the exact answer.

45. $\quad\ \ 51.701$
$\qquad 428.8$
$\underline{+\ \ \ 4.518}$

46. $\quad\ \ 51.6$
$\underline{-\ 17.98}$

47. $\quad\ \ 147.36$
$\qquad 28.701$
$\underline{+\ 56.939}$

48.   269.87   49.   584.786   50.   897.2
       491.6   − 90.27   − 34.718
     + 289.29

First, use front end rounding and estimate each answer. Then add or subtract to find the exact answer for each application problem.

51. Kevin sells industrial rubber goods. On one trip he started from Cleveland and drove 167.6 miles to Detroit, then 270.8 miles to Chicago, and finally 341.3 miles back to Cleveland. Find the total length of his trip.

52. Chuck drove on a five-day vacation trip. On the first day he drove 9.7 hours, 2.6 hours on the second day, 12.2 hours on the third day, 1.8 hours on the fourth day, and 13.4 hours on the fifth day. How many hours did he drive?

53. At one gasoline stop, a car odometer read 87,431.2 miles. At the next stop, it read 88,123.1 miles. How far did the car travel between stops?

54. At the beginning of March, the odometer of Marvin Raymond's company car read 30,197.2 miles. At the end of March, it read 42,672.1 miles. How many miles did Marvin drive during he month?

55. The accountant at a lumber yard found that the payroll for one week was $989.23, utilities were $219.22, advertising was $364.89, and payments to lumber mills were $7976.35. Find the amount spent during the week.

56. On February 1, Noc Nuynh had $1009.24 in her checking account. During the month she deposited a tax refund check of $704.42 and her paycheck of $1258.94. She wrote checks totaling $1498.53 and had $200 transferred to her savings account. Find her checking account balance at the end of the month.

DECIMALS

4.4 Multiplying Decimals

Objective 1 **Multiply decimals.**

Multiply.

1. 0.053	2. 0.682	3. 19.3	4. 96.5
$\times$ 4.3	$\times$ 3.9	$\times$ 4.7	$\times$ 4.6

5. 67.6	6. 906	7. 0.074·0.05	8. 0.0009·0.014
$\times$ 0.023	$\times$ 0.081		

9.   0.00321·0.003

In each of the following, find the amount of money earned on a job by multiplying the number of hours worked and the pay per hour. Round your answer to the nearest cent, if necessary.

10. 27 hours at $6.04 per hour

11. 31.6 hours at $9.83 per hour

Find the cost of each of the following.

12. 16 apples at $0.59 each

13. 7 quarts of oil at $1.05 each

Objective 2   **Estimate the answer when multiplying decimals.**

First use front end rounding and estimate the answer. Then multiply to find the exact answer.

14. 49.7	15. 29.8	16. 58.73	17. 32.53
$\times$ 5.8	$\times$ 3.4	$\times$ 3.72	$\times$ 23.26

18. 76.4	19. 2.99	20. 391.9	21. 27.5
$\times$ 0.57	$\times$ 3.5	$\times$ 7.74	$\times$ 11.2

Solve. If the problem involves money, round to the nearest cent, if necessary.

22. Mary Williams pays $32.96 per month for a television payment. How much will she pay over 15 months?

23. Steve's car payment is $309.56 per month for 48 months. How much will he pay altogether?

24. The Duncan family's state income tax is found by multiplying the family income of $32,906.15 by the decimal 0.064. Find their tax.

25. A recycling center pays $0.142 per pound of aluminum. How much would be paid for 176.3 pounds?

4.4 Mixed Exercises

Multiply.

26.   62.92   27.   46.94   28. 9.4×8.3   29. 0.0053·0.009
       × 0.032   × 0.046

30. 0.05×0.0062

In each of the following, find the amount of money earned on a job by multiplying the number of hours worked and the pay per hour. Round your answer to the nearest cent, if necessary.

31. 35 hours at $5.72 per hour   32. 27.6 hours at $7.21 per hour

Find the cost of each of the following.

33. 12 rolls of film at $1.72 each   34. 25 cans of soda at $0.29 each

First use front end rounding and estimate the answer. Then multiply to find the exact answer.

35.   126.21   36.   315.26
       × 11.23   ×   8.7

Solve. If the problem involves money, round to the nearest cent, if necessary.

37. Hertz charges $29.95 a day for a certain can rental, plus $0.29 per mile. Find the cost of a five-day trip of 1126 miles.

38. A motor home rents for $375 per week plus $0.35 per mile. Find the rental cost for a four-week trip of 3450 miles.

DECIMALS

4.5 Dividing Decimals

Objective 1 Divide a decimal by a whole number.

Divide. Round answers to the nearest thousandth, if necessary.

1. $3\overline{)43.95}$ 2. $6\overline{)10.763}$ 3. $33\overline{)77.847}$ 4. $5\overline{)34.84}$

5. $4\overline{)31.974}$ 6. $7\overline{)19.863}$ 7. $11\overline{)46.98}$ 8. $8\overline{)49.2465}$

9. $3\overline{)62.7}$ 10. $8\overline{)224.63}$ 11. $41\overline{)726.43}$ 12. $54\overline{)895.79}$

Find the cost of each item. Round to the nearest cent.

13. 3 pairs of socks for $5.98   14. 5 pounds of applies for $3.99

Objective 2   Divide a decimal by a decimal.

Divide. Round answers to the nearest thousandth, if necessary.

15. $0.3\overline{)38.84}$ 16. $0.9\overline{)3.4166}$ 17. $0.6\overline{)78.59}$ 18. $0.8\overline{)93.52}$

19. $4.5\overline{)79.468}$ 20. $3.4\overline{)436.05}$ 21. $3.1\overline{)726.43}$ 22. $0.82\overline{)7.83}$

23. $0.52\overline{)34.96}$ 24. $2859.4 \div 0.053$ 25. $0.07 \div 0.00043$

Solve each application problem. Round money answers to the nearest cent, if necessary.

26. Leon Williams drove 542.2 miles on the 16.3 gallons of gas in his Ford Taurus.
     How many miles per gallon did he get? Round to the nearest tenth.

27. Raymond Starr bought 7.4 yards of fabric, paying a total of $26.27. Find the cost
     per yard.

28. To build a barbecue, Diana Jenkins bought 589 bricks, paying $185.70. Find the
     cost per brick.

29. Cerise Montoya is a newspaper distributor. Last week she paid the newspaper
     $261.02 for 1684 copies. Find the cost per copy.

30. Marc Karn earned \$335.50 for 50 hours of work. Find his earnings per hour.

Objective 3   Estimate the answer when dividing decimals.

Decide if each answer is **reasonable** *or* **unreasonable** *by rounding the numbers and estimating the answer.*

31. $49.8 \div 7.1 = 7.014$ 　　　　　　　32. $126.2 \div 11.2 = 11.268$

33. $31.5 \div 8.4 = 37.5$ 　　　　　　　　34. $486.9 \div 5.06 = 962.253$

35. $624.7 \div 19.24 = 32.469$ 　　　　　36. $800.2 \div 40.15 = 19.930$

37. $1092.8 \div 37.92 = 2.882$ 　　　　　38. $1564.9 \div 50.049 = 312.674$

39. $8695.15 \div 98.762 = 880.415$ 　　　40. $6608.04 \div 415.6 = 15.9$

Objective 4   Use the order of operations with decimals.

Simplify by using the order of operations

41. $3.7 + 5.1^2 - 9.4$ 　　　　　　　　42. $3.1^2 - 1.9 + 5.8$

43. $42.92 \div 5.8 \cdot 7.3$ 　　　　　　　44. $55.744 \div 6.4 \cdot 1.9$

45. $18.5 + (37.1 - 29.8) \cdot 10.7$ 　　　46. $58.1 - (17.9 - 15.2) \cdot 1.8$

47. $27.51 - 3.2 \cdot 9.8 \div 1.6$ 　　　　　48. $9.1 - 0.07 \cdot 2.1 \div 0.042$

49. $9.8 \cdot 4.76 + 17.94 \div 2.6$ 　　　　50. $62.699 \div 7.42 + 3.6 \cdot 1.4$

4.5 Mixed Exercises

Divide. Round to the nearest thousandth.

51. $46\overline{)54.703}$ 　　52. $67\overline{)32.946}$ 　　53. $34\overline{)28.724}$ 　　54. $21\overline{)91.34}$

Find the cost of each item. Round to the nearest cent.

55. 49 books for \$154.84 　　　　　　56. 200 pencils for \$10.20

Divide. Round to the nearest thousandth, if necessary.

57. $0.96\overline{)0.831}$

58. $0.006\overline{)0.618}$

59. $0.004\overline{)11.628}$

60. $875.469 \div 5.36$

61. $471.82 \div 3.6$

Solve each application problem.

62. At a record manufacturing company, 400 records cost $389. Find the cost per record. Round to the nearest cent.

63. Nicole Mack pays $24.76 per month on a charge account on which she owes $519.96. How long will it take her to pay off the account?

64. Suppose Chris Batansa pays $65.29 per month to Household Finance. How many months will it take to pay off a loan if $2807.47 is owed?

Decide if each answer is reasonable or unreasonable by rounding the numbers and estimating the answer.

65. $17.56 \div 0.82 = 2.141$

66. $573.056 \div 17.6 = 32.56$

67. $6082.59 \div 31.684 = 19.198$

Simplify by using the order of operations.

68. $2.1 + 4.8^2 - 19.6$

69. $14.3 \cdot 2.7 + 9.7 - 3.4$

70. $62.7 - (16.2 - 11.8) \cdot 2.5$

DECIMALS

4.6 Writing Fractions as Decimals

Objective 1 Write a fraction as a decimal.

Write each fraction or mixed number as a decimal. Round to the nearest thousandth, if necessary.

1. $6\frac{1}{2}$

2. $\dfrac{1}{5}$

3. $2\frac{2}{3}$

4. $\dfrac{1}{8}$

5. $\dfrac{1}{11}$

6. $7\frac{1}{10}$

7. $\dfrac{3}{5}$

8. $\dfrac{7}{8}$

9. $4\frac{1}{9}$

10. $\dfrac{13}{25}$

11. $\dfrac{3}{20}$

12. $31\frac{3}{13}$

Objective 2 Compare the size of fractions and decimals.

Find the smaller of the two given numbers. Write < or > to make a true statement.

13. $\dfrac{5}{8}$ ___ 0.634

14. $\dfrac{2}{5}$ ___ 0.401

15. $\dfrac{1}{5}$ ___ 0.19

16. $\dfrac{2}{3}$ ___ 0.67

17. $\dfrac{3}{5}$ ___ 0.599

18. $\dfrac{5}{6}$ ___ 0.83

Arrange in order from smallest to largest.

19. $\dfrac{3}{11}, \dfrac{1}{3}, 0.29$

20. $\dfrac{8}{9}, 0.88, 0.89$

21. $\dfrac{1}{6}, 0.166, 0.1666$

22. $\dfrac{7}{15}, 0.466, \dfrac{9}{19}$

4.6 Mixed Exercises

Write each fraction or mixed number as a decimal. Round to the nearest thousandth if necessary.

23. $\dfrac{7}{15}$

24. $\dfrac{5}{12}$

25. $\dfrac{11}{18}$

26. $19\frac{17}{24}$

90   BASIC COLLEGE MATHEMATICS Chapter 4

Find the smaller of the two given numbers. Write < or > to make a true statement.

27. $\dfrac{1}{25}$ ___ 0.039   28. $\dfrac{3}{8}$ ___ 0.38   29. $\dfrac{5}{9}$ ___ 0.55   30. $\dfrac{3}{16}$ ___ 0.188

Arrange in order from smallest to largest.

31. $\dfrac{1}{7}, \dfrac{3}{16}, 0.187$ 32. $0.8462, \dfrac{11}{13}, \dfrac{6}{7}$

Chapter 5

RATIO AND PROPORTION

5.1 Ratios

Objective 1   **Write ratios as fractions.**

Write each ratio as a fraction in lowest terms.

1. 7 to 8

2. 18 to 24

3. 76 to 101

4. 30 to 84

5. 95 cents to 125 cents

6. 80 miles to 30 miles

7. $85 to $135

8. 5 men to 20 men

Solve each application problem. Write each ratio as a fraction in lowest terms.

9. Mr. Williams is 42 years old, and his son is 18. Find the ratio of Mr. Williams' age to his son's age.

10. When using Roundup vegetation control, add 128 ounces of water for every 6 ounces of the herbicide. Find the ratio of herbicide to water.

Objective 2   **Solve ratio problems involving decimals or mixed numbers.**

Write each ratio as a fraction in lowest terms.

11. $6\frac{1}{2}$ to 2

12. $4\frac{1}{8}$ to 3

13. 3 to $2\frac{1}{2}$

14. 11 to $2\frac{4}{9}$

15. $1\frac{1}{4}$ to $1\frac{1}{2}$

16. $3\frac{1}{2}$ to $1\frac{3}{4}$

Solve each application problem. Write each ratio as a fraction in lowest terms.

17. One refrigerator holds $3\frac{3}{4}$ cubic feet of food, while another holds 5 cubic feet. Find the ratio of the amount of storage in the first refrigerator to the amount of storage in the second.

18. One car has a $15\frac{1}{2}$ gallon gas tank while another has a 22 gallon gas tank. Find the ratio of the amount the first tank holds to the amount the second tank holds.

For each triangle, find the ratio of the length of the longest side to the length of the shortest side. Write each ratio as a fraction in lowest terms.

19.

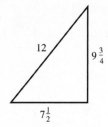

20.

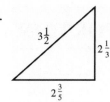

Objective 3   **Solve ratio problems after converting units.**

Write each ratio as a fraction in lowest terms. Be sure to make all necessary conversions.

21. 4 days to 2 weeks   22. 4 feet to 15 inches

23. 6 yards to 10 feet   24. 7 gallons to 8 quarts

25. 40 ounces to 3 pounds   26. 80 cents to $3

Write each ratio as a fraction in lowest terms. Be sure to make all necessary conversions.

27. Find the ratio of $17\frac{1}{2}$ inches to $2\frac{1}{3}$ feet.

28. What is the ratio of $59\frac{1}{2}$ days to $4\frac{1}{4}$ weeks?

5.1 Mixed Exercises

Solve each application problem. Write each ratio as a fraction in lowest terms.

29. The length of a rectangle is 50 inches and the height is 28 inches. Find the ratio of the length of the rectangle to the height of the rectangle.

30. The area of Gettysburg National Military Park is 3900 acres, and the area of Chickamauga and Chattanooga National Military Park is 8100 acres. Find the ratio of the areas of the two parks.

Write each ratio as a fraction in lowest terms.

31. A building is $42\frac{1}{2}$ feet tall. It casts a shadow $1\frac{3}{14}$ feet long. Find the ratio of the height of the building to the length of its shadow.

32. A spruce tree is $28\frac{3}{4}$ feet tall. It casts a shadow $11\frac{1}{2}$ feet long. Find the ratio of the height of the tree to the length of its shadow.

33. $2\frac{1}{3}$ to $2\frac{2}{9}$ 34. $5\frac{1}{2}$ to $8\frac{1}{4}$

Write each ratio as a fraction in lowest terms. Be sure to make all necessary conversions.

35. $3\frac{1}{5}$ inches to 4 feet

36. $4\frac{2}{3}$ days to 1 week

37. $9\frac{1}{3}$ ounces to $3\frac{1}{2}$ pounds

38. $5\frac{1}{2}$ gallons to $2\frac{3}{4}$ quarts

RATIO AND PROPORTION

5.2 Rates

Objective 1 **Write rates as fractions.**

Write each rate as a fraction in lowest terms.

1. 75 miles in 25 minutes

2. 85 feet in 17 seconds

3. 28 dresses for 4 people

4. 70 horses for 14 teams

5. 45 gallons in 3 hours

6. 225 miles on 15 gallons

7. 119 pills for 17 patients

8. 144 kilometers on 16 liters

9. 256 pages for 8 chapters

10. 990 miles in 18 hours

Objective 2 **Find unit rates.**

Find each unit rate.

11. $75 in 5 hours

12. $3500 in 20 days

13. $1540 in 14 days

14. $7875 for 35 pounds

15. $122.76 in 9 hours

16. 189.88 miles on 9.4 gallons

Solve each application problem.

17. Eric can pack 12 crates of berries in 24 minutes. Give his rate in rate per minute and in minutes per crate.

18. Michelle can plow 7 acres in 14 hours. Give her rate in acres per hour and in hours per acre.

19. Candy makes $220.32 in 24 hours. What is her rate per hour?

20. The 4.6 yards of fabric needed for a dress costs $27.14. Find the cost of 1 yard.

Objective 3 **Find the best buy based on cost per unit.**

Find the best buy (based on cost per unit) for each item.

21. Beans: 12 ounces for $1.49, 16 ounces of $1.89

22. Orange juice: 16 ounces for $0.89, 32 ounces for $1.90

23. Cola: 6 cans for $1.98, 12 cans for $3.59, 24 cans for $8

24. Soup: 3 cans for $1.75, 5 cans for $2.75, 8 cans for $4.55

5.2 Mixed Exercises

Find each unit rate.

25. 262.08 miles on 9.6 gallons

26. $2\frac{3}{4}$ pounds for 11 people

27. $96.25 for 11 hours

28. $619.80 for 6 days

Solve each application problem.

29. The cost of 24.3 square yards of carpet is $872.37. Find the cost of 1 square yard.

30. Ms. Jordan bought 145 shares of stock for $1667.50. Find the cost of 1 share.

31. A company pays $3225 in dividends for the 1250 shares of its stock. Find the value of dividends per share.

Find the best buy (based on the cost per unit).

32. Cereal: 10 ounces for $1.34, 15 ounces for $1.76, 20 ounces for $2.29

RATIO AND PROPORTION

5.3 Proportions

Objective 1 Write proportions.

Write each proportion.

1. 11 is to 15 as 22 is to 30.

2. 50 is to 8 as 75 is to 12.

3. 24 is to 30 as 8 is to 10.

4. 36 is to 45 as 8 is to 10.

5. 14 is to 21 as 10 is to 15.

6. 3 is to 33 as 12 is to 132.

7. 26 is to 4 as 39 is to 6.

8. 9 is to 3 as 42 is to 14.

9. $1\frac{1}{2}$ is to 4 as 21 is to 56.

10. $3\frac{2}{3}$ is to 11 as 10 is to 30.

Objective 2 Decide whether proportions are true or false.

*Write each ratio in lowest terms in order to decide if the proportion is **true** or **false**.*

11. $\dfrac{6}{100} = \dfrac{3}{50}$

12. $\dfrac{48}{36} = \dfrac{3}{4}$

13. $\dfrac{3}{8} = \dfrac{21}{28}$

14. $\dfrac{30}{25} = \dfrac{6}{5}$

15. $\dfrac{390}{100} = 27$

16. $\dfrac{35}{21} = \dfrac{3}{4}$

17. $\dfrac{28}{6} = \dfrac{42}{9}$

18. $\dfrac{54}{30} = \dfrac{108}{60}$

19. $\dfrac{15}{24} = \dfrac{25}{35}$

20. $\dfrac{63}{18} = \dfrac{56}{14}$

21. $\dfrac{108}{225} = \dfrac{24}{51}$

22. $\dfrac{40}{70} = \dfrac{28}{49}$

Objective 3 Find cross products.

*Cross multiply to see whether the proportion is **true** or **false**.*

23. $\dfrac{4}{7} = \dfrac{16}{21}$

24. $\dfrac{12}{18} = \dfrac{20}{30}$

25. $\dfrac{10}{45} = \dfrac{6}{27}$

26. $\dfrac{28}{50} = \dfrac{49}{75}$

27. $\dfrac{132}{24} = \dfrac{11}{3}$

28. $\dfrac{210}{300} = \dfrac{14}{20}$

29. $\dfrac{3\frac{1}{2}}{4} = \dfrac{14}{16}$

30. $\dfrac{4\frac{3}{5}}{9} = \dfrac{23}{36}$

31. $\dfrac{21}{28} = \dfrac{5\frac{3}{4}}{7}$

32. $\dfrac{4}{4\frac{2}{3}} = \dfrac{30}{35}$

33. $\dfrac{6\frac{1}{9}}{3\frac{2}{3}} = \dfrac{40}{24}$

34. $\dfrac{22}{54} = \dfrac{6\frac{1}{3}}{5\frac{2}{11}}$

5.3 Mixed Exercises

Write each proportion.

35. 6 is to 21 as 10 is to 35   36. 9 is to 15 as 21 is to 35

Write each ratio in lowest terms in order to decide if the proportion is **true** *or* **false.**

37. $\dfrac{18}{81} = \dfrac{9}{36}$ 38. $\dfrac{52}{100} = \dfrac{39}{75}$ 39. $\dfrac{90}{55} = \dfrac{54}{33}$ 40. $\dfrac{36}{48} = \dfrac{54}{72}$

Cross multiply to see whether the proportion is **true** *or* **false.**

41. $\dfrac{69.9}{3} = \dfrac{100.19}{4.3}$ 42. $\dfrac{3911.0}{403.2} = \dfrac{304.2}{36.4}$

43. $\dfrac{2.98}{7.1} = \dfrac{1.7}{4.3}$ 44. $\dfrac{42.2}{106.8} = \dfrac{84.9}{206}$

RATIO AND PROPORTION

5.4 Solving Proportions

Objective 1   Find the unknown number in a proportion.

Find the unknown number in each proportion.

1. $\dfrac{3}{2} = \dfrac{x}{6}$

2. $\dfrac{9}{4} = \dfrac{36}{x}$

3. $\dfrac{9}{7} = \dfrac{x}{28}$

4. $\dfrac{x}{11} = \dfrac{44}{121}$

5. $\dfrac{35}{x} = \dfrac{5}{3}$

6. $\dfrac{x}{52} = \dfrac{5}{13}$

7. $\dfrac{96}{60} = \dfrac{8}{x}$

8. $\dfrac{7}{5} = \dfrac{98}{x}$

9. $\dfrac{9}{14} = \dfrac{x}{70}$

10. $\dfrac{90}{x} = \dfrac{15}{8}$

11. $\dfrac{x}{110} = \dfrac{7}{10}$

12. $\dfrac{14}{x} = \dfrac{21}{18}$

Objective 2   Find the unknown number in a proportion with mixed numbers or decimals.

Find the unknown number in each proportion. Write the answer as a whole number or a mixed number when possible.

13. $\dfrac{7}{8} = \dfrac{x}{2}$

14. $\dfrac{5}{x} = \dfrac{3}{7}$

15. $\dfrac{x}{6} = \dfrac{4}{9}$

16. $\dfrac{2}{3\frac{1}{4}} = \dfrac{8}{x}$

17. $\dfrac{3}{x} = \dfrac{5}{1\frac{2}{3}}$

18. $\dfrac{x}{6} = \dfrac{5\frac{1}{4}}{7}$

19. $\dfrac{1\frac{1}{5}}{\frac{1}{2}} = \dfrac{6}{x}$

20. $\dfrac{0}{5\frac{1}{3}} = \dfrac{x}{5}$

21. $\dfrac{x}{7\frac{1}{2}} = \dfrac{0}{9\frac{2}{3}}$

22. $\dfrac{3}{x} = \dfrac{0.8}{5.6}$

23. $\dfrac{16}{12} = \dfrac{2}{x}$

24. $\dfrac{4.2}{x} = \dfrac{0.6}{2}$

5.4 Mixed Exercises

Find the unknown number in each proportion. Write the answer as a whole number or a mixed number when possible.

25. $\dfrac{18}{81} = \dfrac{4}{x}$

26. $\dfrac{100}{x} = \dfrac{75}{30}$

27. $\dfrac{125}{75} = \dfrac{x}{33}$

28. $\dfrac{x}{45} = \dfrac{132}{180}$

29. $\dfrac{2\frac{1}{2}}{1\frac{2}{3}} = \dfrac{x}{2}$

30. $\dfrac{2\frac{5}{9}}{x} = \dfrac{23}{\frac{3}{5}}$

31. $\dfrac{10}{x} = \dfrac{2\frac{1}{2}}{2}$ 32. $\dfrac{x}{7.9} = \dfrac{0}{47.4}$ 33. $\dfrac{x}{4.8} = \dfrac{1.5}{1.2}$

RATIO AND PROPORTION

5.5 Solving Application Problems with Proportions

Objective 1 Use proportions to solve application problems.

Set up and solve a proportion for each problem.

1. A gardening service charges $45 to install 50 square feet of sod. Find the charge to install 125 feet.

2. On a road map, a length of 3 inches represents a distance of 8 miles. How many inches represent a distance of 32 miles?

3. If 6 melons cost $9, find the cost of 10 melons.

4. If 22 hats cost $198, find the cost of 12 hats.

5. 6 pounds of grass seed cover 4200 square feet of ground. How many pounds are needed for 5600 square feet.

6. Margie earns $168.48 in 26 hours. How much does she earn in 40 hours?

7. Juan makes $477.40 in 35 hours. How much does he make in 60 hours.

8. If 5 ounces of a medicine must be mixed with 12 ounces of water, how many ounces of medicine would be mixed with 132 ounces of water.

9. The distance between two cities on a road map is 5 inches. The two cities are really 600 miles apart. The distance between two other cities on the map is 8 inches. How many miles apart are these cities?

10. The distance between two cities is 600 miles. On a map the cities are 10 inches apart. Two other cities are 720 miles apart. How many inches apart are they on the map?

11. If 2 visits to a salon cost $80, find the cost of 11 visits.

12. If a 4-minute phone call cots $0.96, find the cost of a 10-minute call.

13. If 150 square yards of carpet cost $3142.50, find the cost of 210 square yards of the carpet.

14. Scott paid $240,000 for a 5-unit apartment house. Find the cost of a 16-unit apartment house.

15. Brian plants his seeds early in the year. To keep them from freezing, he covers the ground with black plastic. A piece with an area of 80 square feet costs $14. Find the cost of a piece with an are of 700 square feet.

16. A taxi ride of 7 miles cots $9.45. Find the cost of a ride of 12 miles.

17. Dog food for 8 dogs cots $15. Find the cost of dog food for 12 dogs.

18. To make battery acid, Jeff mixes $9\frac{1}{2}$ gallons of pure acid with 25 gallons of water. How much acid would be needed for 75 gallons of water?

19. Tax on an $18,000 car is $1620. Find the tax on a $24,000 car.

20. If $18\frac{3}{4}$ yards of material are needed for 5 dresses, how much material is needed for 9 dresses?

Chapter 6

PERCENT

6.1 BASICS OF PERCENT

Objective 1 **Learn the meaning of percent.**

Write as a percent.

1. 43 people out of 100 drive small cars.

2. The tax is $8 per $100.

3. The cost for labor was $45 for every $100 spent to manufacture an item.

4. 29 out of 100 gallons of gas were unleaded.

5. 32 out of 100 students majored in engineering.

6. 38 out of 100 planes departed on time.

7. 52 out of 100 high school students went to college.

Objective 2 **Write percents as decimals.**

Write each percent as a decimal.

8. 37%	9. 42%	10. 83%	11. 310%
12. 510%	13. 9%	14. 4%	15. 10%
16. 32.5%	17. 61.9%	18. 0.025%	19. 0.256%

Objective 3 **Write decimals as percents.**

Write each decimal as a percent.

20. 0.30	21. 0.40	22. 0.2	23. 0.0
24. 0.71	25. 0.86	26. 0.42	27. 0.07
28. 0.09	29. 0.036	30. 0.986	31. 0.564
32. 4.93	33. 3.47	34. 4.2	

Objective 4   **Understand 100%, 200%, and 300%.**

Fill in the blanks.

35. 100% of $19 is _____.

36. 200% of 170 miles is _____.

37. 300% of $76 is _____.

38. 100% of 12 dogs is _____.

Objective 5   **Use 50%, 10%, and 1%.**

Fill in the blanks.

39. 50% of 48 copies is _____.

40. 10% of 4920 televisions is _____.

41. 1% of 400 homes is _____.

42. 50% of 250 signs is _____.

6.1 Mixed Exercises

Write as a percent.

43. The tax is $6.75 per $100. What is the tax rate?

44. 26 out of 100 students are sophomores. What percent are sophomores?

Write each percent as a decimal.

45. 0.302%

46. 0.3%

47. 0.5%

Write each decimal as a percent.

48. 3.4

49. 0.0423

50. 0.0736

51. 4.836

52. 0.0005

Fill in the blanks.

53. 100% of $520 is _____.

54. 300% of 29 days is _____.

55. 50% of 98 hours is _____.

56. 10% of 100 years is _____.

PERCENT

6.2 Percents and Fractions

Objective 1 Write percents as fractions.

Write each percent as a fraction or mixed number in lowest terms.

1. 35%

2. 12%

3. 56%

4. 75%

5. 62.5%

6. 43.6%

7. $16\frac{1}{3}\%$

8. $22\frac{2}{9}\%$

9. $6\frac{2}{3}\%$

10. $46\frac{1}{3}\%$

Objective 2 Write fractions as percents.

Write each fraction or mixed number as a percent. Round percents to the nearest tenth if necessary.

11. $\dfrac{7}{10}$

12. $\dfrac{53}{100}$

13. $\dfrac{81}{100}$

14. $\dfrac{12}{25}$

15. $\dfrac{64}{75}$

16. $\dfrac{33}{50}$

17. $\dfrac{47}{50}$

18. $\dfrac{5}{9}$

19. $\dfrac{4}{7}$

20. $3\frac{4}{5}$

21. $2\frac{3}{4}$

22. $7\frac{2}{5}$

Objective 3 Use the table of percent equivalents.

Complete this chart. Round decimals to the nearest thousandth and percents to the nearest tenth if necessary.

	Fractions	Decimal	Percent
23.	$\frac{1}{2}$	_____	_____
24.	_____	0.125	_____
25.	$\frac{1}{4}$	_____	_____
26.	$\frac{5}{8}$	_____	_____
27.	_____	_____	87.5%

28. $\frac{3}{8}$ _____   _____

29.   _____   _____ $33\frac{1}{3}\%$

30. $\frac{2}{5}$ _____   _____

31.   _____   0.325   _____

32. $\frac{2}{3}$ _____   _____

6.2 Mixed Exercises

Write each percent as a fraction or mixed number in lowest terms.

33. 0.5%   34. 0.9%   35. 140%

36. 175%   37. 225%   38. 0.01%

Write each mixed number as a percent. Round percents to the nearest tenth if necessary.

39. $3\frac{3}{4}$ 40. $4\frac{1}{3}$

Complete this chart. Round decimals to the nearest thousandth and percents to the nearest tenth if necessary.

	Fractions	*Decimal*	*Percent*
41.	_____	_____	$83\frac{1}{3}\%$
42.	$\frac{5}{7}$	_____	_____

PERCENT

6.3 Percents and Fractions

Objective 2 **Solve for an unknown value in a proportion.**

Use the percent proportion $\left(\dfrac{part}{whole} = \dfrac{percent}{100}\right)$ *and solve for the unknown value.*

1. part = 30, percent = 25

2. part = 160, percent = 20

3. part = 7, percent = 40

4. part = 18, percent = 150

5. whole = 48, percent = 25

6. whole = 36, percent = 15

7. whole = 25, percent = 14

8. whole = 50, percent = 17

9. part = 44, whole = 200

10. part = 75, whole = 1500

11. part = 143, whole = 550

12. part = 12, whole = 50

Objective 3 **Identify the percent.**

Identify the percent. Do not try to solve for any unknowns.

13. 35% of 1000 is 350

14. 25% of 750 is 187.5

15. 71% of what number is 438?

16. 83% of what number is 21.5?

17. 36 is 72% of what number?

18. 63 is what percent of 218?

Identify the percent. Do not try to solve for any unknowns.

19. A team won 12 of the 18 games it played. What percent of its games did it win?

20. A chemical is 42% pure. Of 800 grams of the chemical, how much is pure?

21. Sales tax of $8 is charge on an item costing $200. What percent of sales tax is charged?

22. 17% of Tom's check of $340 is withheld. How much is withheld?

Objective 4   **Identify the whole.**

Identify the whole. Do not try to solve for any unknowns.

23. 40% of 48 is 19.2.

24. 16 is 400% of 4.

25. What is 14% of 78?

26. What is 56% of 965?

27. 52 is 12% of what number?

28. 71 is what percent of 384?

29. What percent of 60 is 25?

30. 0.68% of 487 is what number?

Identify the whole in each application problem. Do not try to solve for any unknowns.

31. In one storm, Springbrook got 15% of the season's snowfall. Springbrook's total snowfall for that season was 30 inches. How many inches of snow fell in that one storm?

32. In one state, the sales tax is 8%. On a purchase, the amount of tax was $26. Find the cost of the item purchase.

Objective 5   **Identify the part.**

Identify the part in each application problem. Do not try to solve for any unknowns.

33. 16% of 3500 is 560.

34. 29 is 25% of 116.

35. 29.81 is what percent of 508?

36. What percent of 162 is 85?

37. What number is 12.4% of 1408?

38. 16.74 is 11.9% of what number?

Identify the part in each application problem. Do not try to solve for any unknowns.

39. In a one-day storm, Odentown received 0.3% of the season's total rainfall. Odentown received 4 inches of rain on that day. How many inches of rain fell during the season?

40. A hatchery is notified that 7% of its shipment of baby salmon did not arrive healthy. Of 1500 salmon shipped, how many did not arrive healthy?

41. There are 720 quarts of grape juice in a vat holding a total of 2400 quarts of fruit juice. What percent of the vat is grape juice?

42. A teacher of English literature found that 15% of the students' papers are handed in late. If there are 40 students in a class, how many papers will be handed in late?

6.3 Mixed Exercises

Use the percent proportion $\left(\dfrac{part}{whole} = \dfrac{percent}{100}\right)$ *and solve for the unknown value.*

43. part = 45, percent = 9

44. whole = 400, percent = 16.5

45. part = 2.02, whole = 10.10

Identify the percent.

46. 16% of 20 is 3.2

47. 150% of 300 is 450

48. 90% of 70 is 63

49. 30% of 98 is 29.4

Identify the whole.

50. 20% of 3 is 0.6

51. 15.2% of 200 is 30.4

52. 80% of 40 is 32

Identify the part.

53. 30% of 8 is 2.4

54. 120% of 60 is 72

55. 65% of 1200 is 780

PERCENT

6.4 Using Proportions to Solve Percent Problems

Objective 1 Use the percent proportion to solve for the part.

Use the percent proportion to find the part. Round to the nearest tenth if necessary.

1. 25% of 584   2. 20% of 1400   3. 13% of 270   4. 9% of 42

Use multiplication to find the part. Round to the nearest tenth if necessary.

5. 135% of 35   6. 175% of 50   7. 39.4% of 300   8. 22.5% of 1300

Solve each application problem. Round to the nearest tenth if necessary.

9. A library has 330 visitors of Saturday, 20% of whom are children. How many are children?

10. Bonnie Rae spent 15% of her savings on textbooks. If her savings were $560, find the amount that she spent on textbooks.

11. A survey at an intersection found that of 2200 drivers, 43% were wearing seat belts. How many drivers in the survey were wearing seat belts?

12. A family of four with a monthly income of $2100 spends 90% of its earnings and saves the balance. How much does the family save in one month?

Objective 2 Solve for the whole using the percent proportion.

Use the percent proportion to find the whole. Round to the nearest tenth if necessary.

13. 50 is 10% of what number?   14. 15 is 5% of what number?

15. 36% of what number is 74?   16. 50% of what number is 76?

17. 100 is 25% of what number?   18. 96 is 40% of what number?

19. 548 is 110% of what number?   20. 91 is 130% of what number?

Solve each application problem. Round to the nearest tenth if necessary.

21. On campus this semester there are 2028 married students, which is 26% of the total enrollment. Find the total enrollment.

22. Michael Elders owns stock worth $4250, which is 17% of the value of his investments. What is the value of his investments?

23. There are 18 violin players in an orchestra. If this is 24% of the orchestra membership, find the number of members in the orchestra.

24. This year, there are 960 scholarship applications, which is 120% of the number of applications last year. Find the number of applications last year.

25. At Dee's Sandwich Shop, 20% of the customers order a dill pickle. If 465 dill pickles are sold, find the total number of customers.

26. Kathy Wicklund's overtime pay is $420, which is 12% of her total pay. What is her total pay?

Objective 3   **Find the percent using the percent proportion.**

Use the percent proportion to find the percent. Round to the nearest tenth if necessary.

27. 35 is what percent of 105?

28. 13 is what percent of 50?

29. 15 is what percent of 90?

30. 550 is what percent of 1000?

31. 12 is what percent of 400?

32. 7 is what percent of 280?

33. What percent of 6000 is 12?

34. What percent of 8000 is 4?

35. What percent of 100 is 150?

36. What percent of 72 is 810?

Solve each application problem. Round to the nearest tenth if necessary.

37. In one shipment, 695 out of 27,800 crates were damaged. What percent of the crates were damaged?

38. Total daily circulation of The Sun is 180,000. If complimentary (nonpaid) circulation amounts to 4050 copies per day, what percent of the total circulation is nonpaid circulation?

39. The number of ballots cast in a parish election is 12,969. If the number of registered voters in the parish is 19,800, what percent has voted?

40. Rocky Mountain Water estimates 11,700 gallons of their water will be used in steam irons. If 780,000 gallons are sold, what percent will be used in steam irons?

41. G&G Pharmacy has a total payroll of $89,350, of which $19,657 goes towards employee fringe benefits. What percent of the total payroll goes to fringe benefits?

42. Vera's Antiquery says that of its 5100 items in stock, 4233 are just plain junk, while the rest are antiques. What percent of the number of items in stock is antiques?

6.4 Mixed Exercises

Use the percent proportion to find the part, the whole, or the percent. Round to the nearest tenth if necessary.

43. 0.3% of 515 is what number?

44. 2.1% of what number is 525?

45. What percent of 42 is 10.92?

46. 0.7% of 2100 is what number?

47. 5.25% of what number is 711.9

48. What percent is 3.9 of 4.5?

49. 69 is 4.6% of what number?

50. What number is $81\frac{1}{4}$% of 4032?

51. 156 is 5.2% of what number?

52. 650 is what percent of 13?

53. What number is 0.7% of 3500?

54. 750 is what percent of 25?

Solve each application problem. Round to the nearest tenth if necessary.

55. In one chemistry class, 60% of the students passed? If 90 students passed, how many students were in the class?

56. In the last election, 74% of the eligible people actually voted. If there were 7844 voters, how many people were eligible?

57. This month's class goal for Easy Writer Pen Company is 1,844,500 ballpoint pens. If 239,785 pens have been sold, what percent of the goal has been reached?

58. In a motor cross, the leader has completed 108.8 miles of the 128-mile course. What percent of the total course has she completed?

PERCENT

6.5 Using the Percent Equation

Objective 1 Use the percent equation to solve for the part.

Find the missing part using the percent equation. Round to the nearest tenth if necessary.

1. 40% of 480

2. 70% of 920

3. 65% of 1300

4. 99% of 300

5. 16% of 520

6. 22% of 960

7. 9% of 240

8. 5% of 450

9. 125% of 76

10. 140% of 76

11. 12.4% of 8100

12. 15.3% of 1020

Solve each application problem. Round to the nearest tenth or cent if necessary.

13. A gardener has 56 clients, 25% of whom are residential. Find the number that are residential.

14. The total in sales at Hill's Market last month was $87,428. If the profit was $1\frac{1}{2}$ % of the sales, how much was the profit?

Objective 2 Solve for the whole using the percent equation.

Find the whole using the percent equation. Round to the nearest tenth if necessary.

15. 80 is 20% of what number?

16. 64 is 40% of what number?

17. 50% of what number is 47?

18. 75% of what number is 1125?

19. 540 is 30% of what number?

20. 170 is 20% of what number?

21. $12\frac{1}{2}$ % of what number is 270?

22. $6\frac{1}{4}$ % of what number is 75?

23. $1\frac{1}{4}$ % of what number is 11.25?

24. $2\frac{1}{2}$ % of what number is 15?

25. 200% of what number is 30?

26. 170% of what number is 1462?

Solve each application problem.

27. A tank of an industrial chemical is 25% full. The tank now contains 160 gallons. How many gallons will it contain when it is full?

28. Greg has completed 37.5% of the units needed for a degree. If he has completed 45 units, how many are needed for a degree?

Objective 3 **Find the percent using the percent equation.**

Find the percent using the percent equation. Round to the nearest tenth if necessary.

29. 20 is what percent of 40?

30. 15 is what percent of 75?

31. 13 is what percent of 25?

32. 72 is what percent of 400?

33. What percent of 140 is 49?

34. What percent of 250 is 112.5?

35. What percent is 1.35 of 90?

36. What percent of 160 is 8?

37. 307.2 is what percent of 960?

38. 330 is what percent of 750?

39. 450 is what percent of 200?

40. 125 is what percent of 75?

Solve each application problem.

41. The Robinson family earns $2800 per month and saves $700 per month. What percent of the income is saved?

42. The Hogan family drove 145 miles of their 500-mile vacation. What percent of the total number of miles did they drive?

6.5 Mixed Exercises

Use the percent equation to find the whole, the percent, or the part. Round to the nearest tenth if necessary.

43. What number is 0.4% of 350?

44. 300% of what number is 96?

45. What percent of 15 is 45?

46. 35 is 153% of what number?

47. 0.6% of 510 is what number?

48. What percent of 27 is 90?

49. $105\frac{1}{2}\%$ of what number is 2110?

50. What number is 0.2% of 480?

51. What percent of 36 is 20?

52. 15 is 125% of what number?

53. What percent of 18 is 44.1?

54. 6.75% of 31,252 is what number?

PERCENT

6.6 Solving Application Problems with Percent

Objective 1 Find sales tax.

Find the amount of sales tax and the total cost.

	Amount of Sale	Tax Rate			Amount of Sale	Tax Rate
1.	$100	3%		2.	$200	6%
3.	$50	7%		4.	$170	9%
5.	$215	5%		6.	$15	2%
7.	$30	8%		8.	$78	1%
9.	$67	9%		10.	$450	7%

Solve the following application problems.

11. If the sales tax is 6.5% and the sales are $350, find the amount of sales tax.

12. Find the sales tax on a car costing $14,900 if the sales tax rate is 6.25%.

Objective 2 Find commissions.

Find the commission earned. Round to the nearest cent if necessary.

	Sales	Rate of Commission			Sales	Rate of Commission
13.	$100	15%		14.	$500	8%
15.	$1000	11%		16.	$5783	4%
17.	$3200	32%		18.	$1500	15%
19.	$6225	2.5%		20.	$75,000	4%
21.	$156,000	3%		22.	$25,000	15%

Objective 3   **Find the discount and sales price.**

Find the amount of discount and the amount paid after the discount. Round to the nearest cent if necessary.

	Original Price	Rate of Discount			Original Price	Rate of Discount
23.	$100	25%		24.	$200	15%
25.	$780	10%		26.	$38	40%
27.	$17.50	50%		28.	$22.50	30%
29.	$125	35%		30.	$24.95	60%
31.	$595.80	20%		32.	$205.50	5%

Objective 4   **Find the percent of change.**

Solve each application problem. Round to the nearest tenth of a percent if necessary.

33. Enrollment in secondary education courses increased from 1900 students last semester to 2280 students this semester. Find the percent of increase.

34. The number of days employees of Prodex Manufacturing Company were absent from their jobs decreased from 96 days last month to 72 days this month. Find the percent of decrease.

35. In the past year, the average price of regular unleaded gasoline increased from $1.33 per gallon to $1.69 per gallon in the state of Pennsylvania. Find the percent of increase.

36. The earnings per share of Amy's Cosmetic Company decreased from $1.20 to $0.86 in the last year. Find the percent of decrease.

37. The membership of Pleasant Acres Golf Club was 320 two years ago. The membership is now 740. Find the percent of increase in the two years.

38. The price of a certain model of calculator was $33.50 five years ago. This calculator now costs $18.75. Find the percent of decrease in the price in the last five years.

39. Joe Kooima's income this year was $32,500. Last year his income was $30,000. Find the percent of increase of his income.

40. Last year John Rivera planted 25 acres of corn. This year he planted 32 acres of corn. Find the percent of increase in corn acreage.

6.6 Mixed Exercises

Solve the following application problems. Round money answers to the nearest cent and rates to the nearest tenth of a percent if necessary.

41. In one day, Emma's Boutique sold $945 worth of items. If the sales tax is 6%, find the total amount of money received that day.

42. A television set sells for $750 plus 8% sales tax. Find the price of the set including sales tax.

43. A gold bracelet costs $1300 not including a sales tax of $71.50. Find the rate of sales tax.

44. The sales tax on a book was $1.44. If the sales tax rate is 6%, find the cost of the book.

45. Nicole Sutorius has sales of $18,306 in the month of October. If her rate of commission is 10%, find the amount of commission that she earned.

46. A salesperson at the Sewing Mart was paid $330 in commissions on sales of $5500. Find the rate of commission.

47. Solomon Grundy's has just been sold for $1,692,804. The real estate agent selling the property earned a commission of $42,320.10. Find the rate of commission.

48. A real estate agent sells a house for $135,000. A sales commission of 6% is charged. The agent gets 55% of this commission. How much does the agent get?

49. Geishe's Shoes sells shoes at 33% off the regular price. Find the price of a pair of shoes normally priced at $54, after the discount is given.

50. Mike Lee can purchase a new car at 8% below window sticker price. Find the amount he can save on a car with a window sticker price of $17,608.

51. A "Super 35% Off Sale" begins today. What is the price of a hair dryer normally priced at $15?

52. What is the sales price of a bedroom set priced at $1195, with a discount of 30%?

53. Find the total price of a boat with an original price of $15,000 if it is sold at an 18% discount. Sales tax is 4.75%.

54. A compact disc player normally priced at $450 is on sale for 25% off. Find the discount and the sale price.

55. In 1955, there were 4,097,000 births in the U.S. In 1987, there were 3,829,000 births in the U.S. Find the percent decrease.

56. The population of black squirrels in Earlville, Ohio, has increased from 24 to 130 in the last twenty years. Find the percent of increase.

PERCENT

6.7 Simple Interest

Objective 1   Find the simple interest on a loan.

Find the interest.

	Principle	Rate	Time in Years			Principle	Rate	Time in Years
1.	$200	10%	1		2.	$400	2%	3
3.	$300	12%	4		4.	$1000	12%	2
5.	$80	5%	1		6.	$175	13%	2
7.	$1500	3%	6		8.	$5280	8%	5

	Principle	Rate	Time in Months			Principle	Rate	Time in Months
9.	$200	16%	3		10.	$400	9%	6
11.	$500	11%	9		12.	$1000	12%	12
13.	$820	3%	18		14.	$92	9%	10
15.	$780	6%	5		16.	$522	8%	9

Solve the following application problems.

17. Debbie Ondrika deposits $680 at 14% for 1 year. How much interest will she earn?

18. Bugby Pest Control invests $1500 at 16% for 6 months. What amount of interest will the company earn?

Objective 2   Find the total amount due on a loan.

Find the total amount due on the following loans. *Round to the nearest cent if necessary.*

	Principle	Rate	Time			Principle	Rate	Time
19.	$200	11%	1 year		20.	$3000	5%	6 months
21.	$540	12%	3 months		22.	$1020	10%	2 years

	Principle	Rate	Time		Principle	Rate	Time
23.	$1500	8%	18 months	24.	$5000	7%	5 months
25.	$2210	9%	6 months	26.	$5820	6%	1 year

Solve the following application problems. Round to the nearest cent if necessary.

27. Mary Ann borrows $1200 at 10% for 3 months. Find the total amount due.

28. An investor deposits $7000 at 16% for 2 years. If there are no withdrawals or further deposits, find the total amount in the account after 2 years.

6.7 Mixed Exercises

Find the interest. Round to the nearest cent if necessary.

	Principle	Rate	Time		Principle	Rate	Time
29.	$780	10%	$2\frac{1}{2}$ years	30.	$360	6%	$1\frac{1}{2}$ years
31.	$620	16%	$1\frac{1}{4}$ years	32.	$2000	12%	$3\frac{1}{4}$ years
33.	$650	9%	6 months	34.	$2480	14%	6 months
35.	$14,500	7%	7 months	36.	$10,800	8%	5 months

Find the total amount due on the following loans. Round to the nearest cent if necessary.

	Principle	Rate	Time		Principle	Rate	Time
37.	$1780	12%	6 months	38.	$15,400	16%	5 years
39.	$18,200	7%	8 months	40.	$22,400	9%	6 months

Solve the following application problems. Round to the nearest cent if necessary.

41. Diane DeGroot lends $6500 for 18 months at 12%. How much interest will she earn?

42. A mother lends $6500 to her daughter for 6 months and charges 9% interest. Find the interest charged on the loan.

43. A loan of $1500 will be paid back with 12% interest at the end of 27 months. Find the total amount due.

44. An employee credit union pays 7% interest. If Mario deposits $2100 in his account for $\frac{1}{3}$ year and makes no withdrawals or further deposits, find the total amount in Mario's account after that time.

PERCENT

6.8 Compound Interest

Objective 3 **Solve for the compound amount.**

Find the compound amount given the following deposits. Interest is compounded annually. Round to the nearest cent if necessary.

1. $7000 at 5% for 3 years

2. $2500 at 8% for 4 years

3. $1200 at 2% for 3 years

4. $3200 at 7% for 2 years

5. $4500 at 4% for 2 years

6. $7000 at 6% for 3 years

Objective 4 **Use a compound interest table.**

Use the table for compound interest in your textbook to find the compound amount. Interest is compounded annually. Round to the nearest cent if necessary.

7. $1000 at 6% for 4 years

8. $1000 at 8% for 8 years

9. $4000 at 5% for 9 years

10. $7500 at 6% for 7 years

11. $60 at 5.5% for 2 years

12. $48 at 8% for 3 years

13. $37.50 at 8% for 3 years

14. $71.95 at 5% for 5 years

15. $8428.17 at $4\frac{1}{2}$% for 6 years

16. $10,422.75 at $5\frac{1}{2}$% for 12 years

Objective 5 **Find the compound amount and the amount of compound interest.**

Find the missing amounts for interest compounded annually. Round to the nearest cent if necessary. Use the table for compound interest in your textbook to find the compound amount.

	Principle	Rate	Time in Years	Compound Amount	Compound Interest
17.	$1000	5%	10	_____	_____
18.	$1000	$3\frac{1}{2}$%	7	_____	_____
19.	$8500	6%	12	_____	_____
20.	$12,800	$5\frac{1}{2}$%	9	_____	_____

	Principle	Rate	Time in Years	Compound Amount	Compound Interest
21.	$9150	8%	8	_____	_____
22.	$45,000	4%	4	_____	_____
23.	$21,400	$4\frac{1}{2}$%	11	_____	_____
24.	$78,000	3%	12	_____	_____

6.8 Mixed Exercises

Find the compound amount. Interest is compounded annually. Round to the nearest cent if necessary. Use the table for compound interest in your textbook to find the compound amount.

25. $8000 at 8% for 12 years

26. $3200 at 6% for 5 years

27. $1000 at 4.5% for 6 years

28. $24,600 at 5% for 4 years

29. $35,230 at 3% for 6 years

Solve each application problem. Use the table for compound interest in your textbook to find the compound amount.

30. Scott Williams lends $9000 to the owner of a new restaurant. He will be repaid at the end of 6 years at 8% interest compounded annually. Find how much he will be repaid and how much interest he will earn.

31. Michelle Roberts invests $2500 in a health spa. She will be repaid at the end of 5 years at 6% interest compounded annually. Find how much she will be repaid and how much interest she will earn.

Chapter 7

MEASUREMENT

7.1 Problem Solving with English Measurement

Objective 1   Learn the basic measurement units in the English system.

Fill in the blanks with measurement relationships you have memorized.

1.   1 ft = _____ in

2.   _____ oz = 1 lb

3.   _____ pt = 1 qt

4.   1 T = _____ lb

5.   1 mi = _____ ft

6.   _____ qt = 1 gal

7.   1 c = _____ fl oz

8.   1 yd = _____ ft

9.   _____ hr = 1 day

10. 60 min = _____ hr

Objective 2 Convert among measurement units using multiplication or division.

Convert each measurement using multiplication or division.

11. 72 in to feet

12. 4 qt to gallons

13. 24 fl oz to cups

14. 2 days to hours

15. 10,560 ft to miles

16. 7 yd to feet

17. 24 c to pints

18. $2\frac{1}{2}$ gal to quarts

19. 9000 sec to minutes

20. 3 mi to feet

Objective 3   Convert among measurement units using unit fractions.

Convert these measurements using unit fractions.

21. 12 ft to yards

22. 28 pt to gallons

23. 12 qt to gallons

24. 60 oz to pounds

25. 3000 lb to tons

26. 4 mi to feet

27. 30 in to yards

28. 38 c to pints

29. 5 days to hours

30. 3 hr to minutes

<u>Objective 4</u>   **Solve application problems using English measurement.**

Solve each application problem using the six problem-solving steps.

31. Lee paid $4.65 for 14 oz of honey baked ham. What is the price per pound, to the nearest cent?

32. Sweet Suzies Shop makes 49,000 lb of fudge each week.

(a) How many tons of fudge are produced each day of the 7-day week?

(b) How many tons of fudge are produced each year?

7.1 Mixed Exercises

Convert each measurement.

33. 5 T to pounds

34. 40 pt to gallons

35. $3\frac{1}{2}$ lb to ounces

36. $3\frac{1}{4}$ gal to quarts

37. $17\frac{1}{2}$ ft to inches

38. $2\frac{1}{2}$ yd to feet

39. 60 days to weeks

40. 27,720 ft to miles

41. 75 sec to minutes

42. 380 min to hours

Solve each application problem using the six problem-solving steps.

43. Clarissa paid $1.79 for 4.5 oz of nuts. What is the cost per pound, to the nearest cent?

44. At a preschool, each of 20 children drinks about $\frac{3}{4}$ c of juice with their snack each day. The school is open 3 days a week. How many quarts of juice are needed for 2 weeks of snacks?

MEASUREMENT

7.2 The Metric System – Length

Objective 2 Use unit fractions to convert among units.

Convert each measurement using unit fractions.

1. 1 m to millimeters

2. 1 m to kilometers

3. 1 m to centimeters

4. 1 mm to meters

5. 1 cm to meters

6. 1 km to meters

7. 7 m to millimeters

8. 25.87 m to centimeters

9. 2.3 m to millimeters

10. 53.1 m to centimeters

Objective 3 Move the decimal point to convert among units.

Convert each measurement. Use the metric conversion line.

11. 63.6 cm to meters

12. 807 mm to meters

13. 9600 cm to meters

14. 140 mm to meters

15. 229.7 cm to millimeters

16. 1.94 cm to millimeters

17. 6.4 km to meters

18. 10.35 km to meters

19. 14,500 m to kilometers

20. 25,693 m to kilometers

7.2 Mixed Exercises

Convert each measurement.

21. 3.5 km to centimeters

22. 2.86 km to centimeters

Solve each application problem.

23. Leon's waist size is 72 cm. Give his waist size in millimeters.

24. A dog is 602 mm in length. Give its length in centimeters.

25. A driver is told to turn left in 0.61 km. How many meters is this?

26. A building is 83.6 m tall. How many kilometers is this?

27. Is 103 cm more or less than 1 m?

28. Is 4.72 m more or less than 271 cm?

29. Is 5.38 m more or less than 5000 mm?

30. Is 23 km more or less than 2311 m?

MEASUREMENT

7.3 The Metric System – Capacity and Weight (Mass)

Objective 2 Convert among metric capacity units.

Convert each measurement. Use unit fractions or the metric conversion line.

1. 7 L to kiloliters

2. 9.7 L to milliliters

3. 2.5 L to milliliters

4. 32.4 kL to milliliters

5. 836 kL to liters

6. 523 mL to liters

7. 7863 mL to liters

8. 7724 mL to kiloliters

Objective 4 Convert among metric weight (mass) units.

Convert each measurement. Use unit fractions or the metric conversion line.

9. 9000 g to kilograms

10. 27,000 g to kilograms

11. 6.3 kg to grams

12. 0.76 kg to grams

13. 4.7 g to milligrams

14. 4.91 kg to milligrams

15. 8745 mg to kilograms

16. 42 mg to grams

Objective 5 Distinguish among basic metric units of lengths, capacity, and weight
(mass).

*Write the most appropriate metric unit in each blank. Choose from **km, m, cm, mm, L,
mL, kg, g,** and **mg.***

17. The tablet contains 200 _____ of aspirin.

18. Buy a 5 _____ bottle of water.

19. She drove 450 _____ in one day.

20. The piece of wood is 20 _____ wide.

7.2 Mixed Exercises

Convert each measurement.

21. 9.75 kg to grams

22. 8.71 L to kiloliters

Today medical measurements are usually given in the metric system. Since we convert among metric units of measure by moving the decimal point, it is possible that mistakes can be made. Examine the following dosages and indicate whether they are **reasonable** *or* **unreasonable.**

23. Drink 1.9 kL of Kaopectate after each meal.

24. Apply 0.6 mL of ear drops three times a day.

25. Inject 3 L of insulin each morning.

26. Take 20 mL of cough syrup every four hours.

Write the most appropriate metric unit in each blank. Choose from **km, m, cm, mm, L, mL, kg, g,** *and* **mg.**

27. The doctor administered a 40 _____ injection.

28. The room is 7 _____ wide.

MEASUREMENT

7.4 Problem Solving with Metric Measurement

Objective 1 Solve application problems involving metric measurement.

Solve each application problem.

1. How many 150 mL servings are in 9 L of juice?

2. A commodity costs $0.85 per kilogram. Find the cost of 3 kg 70 g. Round your answer to the nearest cent.

3. Metal chain costs $5.26 per meter. Find the cost of 2 m 47 cm of the chain. Round your answer to the nearest cent.

4. In a laboratory each mouse requires 25 g of food per day. How many kilograms of food are needed to feed 357 mice each day?

5. A 70-L drum is filled with oil which is to be packaged into 140-mL bottles. How many bottles can be filled.?

6. A patio garden slab measures 50 cm by 50 cm by 5 cm and weighs 80 kg. How many kilograms would a truck load of 70 slabs weight?

7. Helene has 3 m 47 cm of red fabric left from one project and 4 m 86 cm of the same fabric from another project. Find the total amount of red fabric she has left in meters.

8. A boy weighed 4 kg 82 g at birth. A month later he weighted 6 kg 17 g. How much weight had he gained, in kilograms?

9. If you drink 175 mL of soda pop every day for two weeks, how many liters would you consume in this time period?

10. If 1.8 kg of candy is to be divided equally among 9 children, how many grams will each child receive?

MEASUREMENT

7.5 Metric-English Conversions and Temperature

Objective 1 Use unit fractions to convert between metric and English units.

Use the table in your textbook and unit fractions to make the following conversions. Round the answer to the nearest tenth.

1. 30 m to yards

2. 12.9 m to feet

3. 1.9 m to inches

4. 14.2 km to miles

5. 9.68 kg to pounds

6. 46.8 L to quarts

7. 50 yd to meters

8. 12 ft to meters

9. 580 yd to meters

10. 5.2 yd to meters

11. 40.2 lb to kilograms

12. 14.6 gal to liters

Solve each application problem. Round each problem to the nearest tenth unless otherwise indicated.

13. Suppose you decide to put together a do-it-yourself picture frame that measures 24 cm by 30 cm. The wood for the frame costs $1.40 per foot. Find the approximate cost of the wood. Round your answer to the nearest cent. (Hint: Start by finding the number of meters of wood in the frame.)

14. A recipe calls for 2.5 L of chicken broth. How many quarts of chicken broth should be used to make this recipe?

15. Geraldo weighs 163 lb. Find his weight in kilograms.

16. The distance from Pittsburgh to Philadelphia is 291 mi. Find the distance in kilometers.

Objective 3 Convert temperature by using the order of operations.

Use the table in your textbook and unit fractions to make the following conversions. Round the answer to the nearest tenth.

17. 62°F

18. 95°F

19. 113°F

20. 446°F

21. 80°F

22. 125°F

23. 10°C   24. 30°C   25. 0°C

26. 60°C   27. 100°C   28. 150°C

Solve each application problem. Round to the nearest degree if necessary.

29. The highest temperature during a recent year in Phoenix was 115°F. Convert this temperature to Celsius.

30. A recipe for roast beef calls for an oven temperature of 200°C. Convert this to Fahrenheit.

7.5 Mixed Exercises

Use the table in your textbook to make the following conversions. Round the answer to the nearest tenth if necessary.

31. 50.4 L to quarts   32. 875 g to pounds

33. 20 L to gallons   34. 108 km to miles

35. 12 qt to liters   36. 96 in to meters

37. 72 mi to kilometers   38. 175 lb to kilograms

Solve each application problem.

39. If gasoline sells for $1.29 per gal, find the cost of one liter. Round to the nearest cent.

40. If paint sells for $11 per gal, find the cost of four liters.

41. Cheryl is making a dress for each of her twin nieces. Each dress requires 72 in of lace trimming. If the lace costs $1.50 per meter, how much will the lace cost to the nearest cent?

42. A 3-L bottle of beverage sells for $2.90. A gallon bottle of the same beverage sells for $3.60. Which is the better value?

43. A kiln for firing pottery reaches a temperature of 450°C. Convert this to Fahrenheit.

44. The temperature of the water in a lake in December is 40°F. Convert this to Celsius. Round to the nearest degree.

45. The average temperature of Odentown was 33 °C during August. Convert to Fahrenheit. Round to the nearest degree.

46. The temperature of a pool in early fall is 52 °F. Convert to Celsius. Round to the nearest degree.

Chapter 8

GEOMETRY

8.1 Basic Geometry Terms

Objective 1 Identify lines, line segments, and rays.

Name each line, line segment, or ray using the appropriate symbol.

1.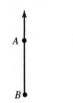

2.

3.

4.

5.

6.

7.

8.

9.

10.

Objective 2 Identify parallel and intersecting lines.

*Label each pair of lines as **parallel** or **intersecting**.*

11.

12.

13.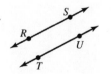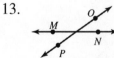

14.

15.

16.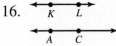

17.

18.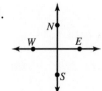

19.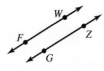

20.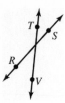

Objective 3 **Identify and name an angle.**

Name each angle that is drawn with darker rays by using the three-letter form of identification.

21.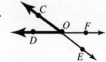

22.

23.

24.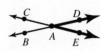

25.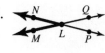

26.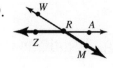

27.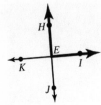

28.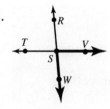

29.

30.

Objective 4 Classify an angle as right, acute, straight, or obtuse.

Label each of the following angles as **acute, right, obtuse,** *or* **straight.**

31.

32.

33.

34.

35.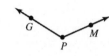

36.

37.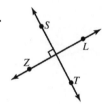

38.

39.

40.

Objective 5 Identify perpendicular lines.

Label each pair of lines as **perpendicular, parallel,** *or* **intersecting.**

41.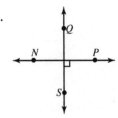

42.

43.

44.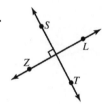

45.

46.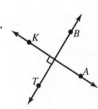

47.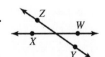

48.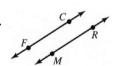

49.

50.

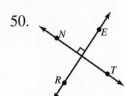

8.1 Mixed Exercises

Name each **line,** **line segment,** *or* **ray** *using the appropriate symbol.*

51.

52.

53.

54.

Label each pair of lines as **parallel** *or* **intersecting.**

55.

56.

57.

58.

Label each angle that is drawn with darker rays by using the three-letter form of identification.

59.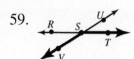

60.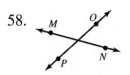

61.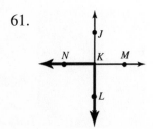

62.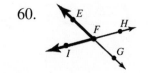

Label each of the following angles as **acute, right, obtuse,** *or* **straight.**

63.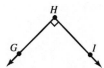

64.

65.

66.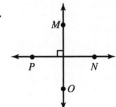

Label each pair of lines as **perpendicular, parallel,** *or* **intersecting.**

67.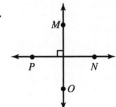

68.

69.

70.

GEOMETRY

8.2 Angles and Their Relationship

Objective 1 Identify complementary angles and supplementary angles.

Find the complement of each angle.

1. 12° 2. 43° 3. 72° 4. 66°

Find the supplement of each angle.

5. 121° 6. 16° 7. 168° 8. 38°

Identify each pair of complementary angles.

9.

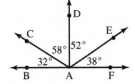

10.

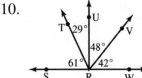

Identify each pair of supplementary angles.

11.

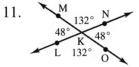

12.

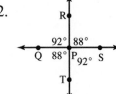

Objective 2 Identify congruent angles and vertical angles.

In each of the following, identify the angles that are congruent.

13.

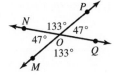

14.

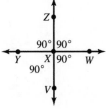

15.

16.

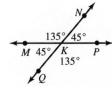

17.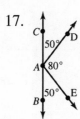

18.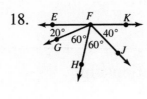

In this figure, ∠AGF measures 42° and ∠BGC measures 105°. Find the number of degrees in each angle.

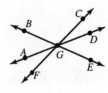

19. ∠CGD   20. ∠EGF   21. ∠DGE   22. ∠BGA

8.2 Mixed Exercises

Find the complement of each angle.

23. 27°   24. 62°   25. 89°   26. 15°

27. 42°   28. 6°

Find the supplement of each angle.

29. 90°   30. 45°   31. 67°   32. 142°

33. 100°   34. 122°   35. 136°   36. 175°

In this figure ∠APB measures 93°, ∠BPC measures 37°. Find the number of degrees in each angle.

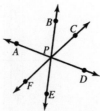

37. ∠CPD   38. ∠DPE   39. ∠EPF   40. ∠FPA

GEOMETRY

8.3 Rectangles and Squares

Objective 1 **Find the perimeter and area of a rectangle.**

Find the perimeter and area of each rectangle.

1. 4 centimeters by 8 centimeters

2. 17 inches by 12 inches

3. 1 centimeter by 17 centimeters

4. 14.5 meters by 3.2 meters

5. $4\frac{1}{2}$ yards by $6\frac{1}{2}$ yards

6. 87.2 feet by 33 feet

7. 37.4 centimeters by 103.2 centimeters

Solve each application problem.

8. A picture frame measures 20 inches by 30 inches. Find the perimeter and area of the frame.

9. A lot is 114 feet by 212 feet. County rules require that nothing be built on land within 12 feet of any edge of the lot. Find the area on which you cannot build.

10. A room is 14 yards by 18 yards. Find the cost to carpet this room if carpet costs $23 per square yard.

Objective 2 **Find the perimeter and area of a square.**

Find the perimeter and area of each square.

11. 9 meters by 9 meters

12. A square 9.2 yards wide

13. A square 7.8 feet wide

14. 13 feet by 13 feet

15. $1\frac{2}{5}$ inches by $1\frac{2}{5}$ inches

16. 8.2 kilometers by 8.2 kilometers

17. 3.1 centimeters by 3.1 centimeters

18. 7.4 inches on each side

19. $4\frac{2}{3}$ miles by $4\frac{2}{3}$ miles

20. 21 meters by 21 meters

Objective 3 **Find the perimeter and area of a composite shape.**

Find the perimeter and area of each figure. All angles that appear to be right angles are indeed right angles.

21.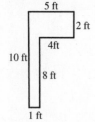

22.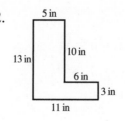

23.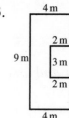

24.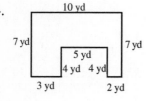

25.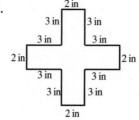

26.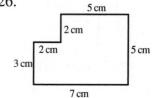

27.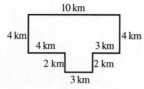

28.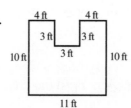

29.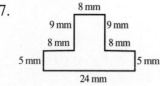

30.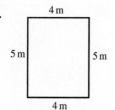

8.3 Mixed Exercises

Find the **perimeter** *and* **area** *of each rectangle.*

31.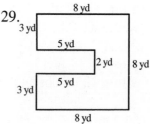

32.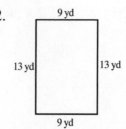

33.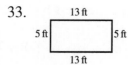

Find the perimeter and area of each square.

34.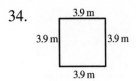

35.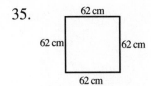

36.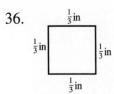

Find the perimeter and area of each figure. All angles that appear to be right angles are indeed right angles.

37.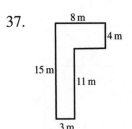

38.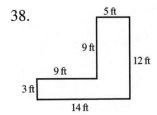

39.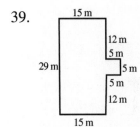

40.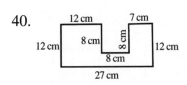

41.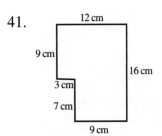

42.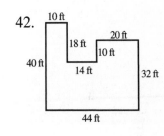

GEOMETRY

8.4 Parallelograms and Trapezoids

Objective 1 Find the perimeter and area of a parallelogram.

Find the perimeter of each parallelogram.

1.

48 m
36 m   36 m
48 m

2.

10.1 in
6.3 in   6.3 in
10.1 in

3.

$7\frac{1}{2}$ ft
$7\frac{1}{2}$ ft
$7\frac{1}{2}$ ft
$7\frac{1}{2}$ ft

Find the area of each parallelogram.

4.

23 yd
31 yd

5.

9.8 m
12.6 m

6.

$2\frac{1}{2}$ m
$4\frac{1}{2}$ m

Solve each application problem.

7. A parallelogram has a height of 3.2 meters and a base of 4.6 meters. Find the area.

8. A parallelogram has a height of $15\frac{1}{2}$ feet and a base of 20 feet. Find the area.

9. A swimming pool is in the shape of a parallelogram with a height of 9.6 meters and base of 12 meters. Find the cost of a solar pool cover that sells for $5.10 per square meter.

10. An auditorium stage has a hardwood floor that is shaped like a parallelogram, having a height of 30 feet and a base of 40 feet. If a company charges $0.65 per square foot to refinish floors, find the cost of refinishing the stage floor.

Objective 2 Find the perimeter and area of a trapezoid.

Find the perimeter of each trapezoid.

11.

42 in
26 in   30 in
61 in

12.

276.2
78.6 cm   61 cm
293 cm

13.

$9\frac{1}{2}$ yd
$15\frac{1}{2}$ yd $18\frac{1}{2}$ yd
$10\frac{1}{4}$ yd

Find the area of each figure.

14.
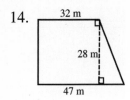
32 m
28 m
47 m

15.
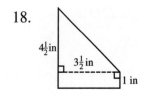
50.3 cm
31 cm
75.1 cm

16.
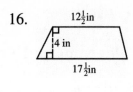
12½ in
4 in
17½ in

17.
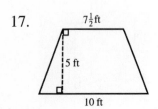
7½ ft
5 ft
10 ft

18.
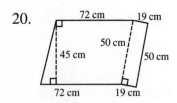
4½ in
3½ in
1 in

19.
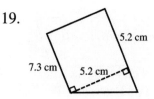
5.2 cm
7.3 cm
5.2 cm

20.
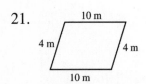
72 cm   19 cm
50 cm
45 cm   50 cm
72 cm   19 cm

8.4 Mixed Exercises

*Find the **perimeter** of each parallelogram.*

21.
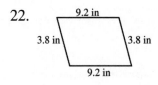
10 m
4 m   4 m
10 m

22.
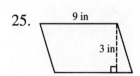
9.2 in
3.8 in   3.8 in
9.2 in

23.
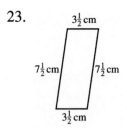
3½ cm
7½ cm   7½ cm
3½ cm

Find the area of each parallelogram.

24.
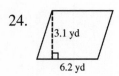
3.1 yd
6.2 yd

25.
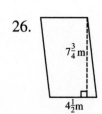
9 in
3 in

26.
7¾ m
4½ m

Find the area of each figure.

27.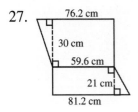

28.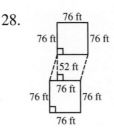

Solve each application problem.

29. The lobby in a resort hotel is in the shape of a trapezoid. The height of the trapezoid is 52 feet and the bases are 47 feet and 59 feet. Carpet that costs $2.75 per square foot is to be laid in the lobby. Find the cost of the carpet.

30. The backyard of a new home is shaped like a trapezoid, having a height of 35 feet and bases of 90 feet and 110 feet. Find the cost of planting a lawn in the yard if the landscaper charges $0.20 per square foot.

31. A hot tub is in the shape below. Find the cost of a cover for the hot tub at a cost of $9.70 per square foot. Angles that appear to be right angles are indeed right angles.

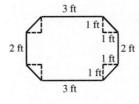

GEOMETRY

8.5 Triangles

Objective 1 Find the perimeter of a triangle.

Find the perimeter of each triangle.

1.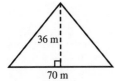
8 yd 6 yd 11 yd

2.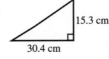
25.7 cm 13.7 cm 19.6 cm

3.
12 yd 15 yd 7 yd

4.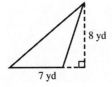
$1\frac{1}{2}$ m $1\frac{1}{3}$ m $2\frac{1}{6}$ m

5.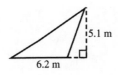
12.4 ft 12.4 ft 12.4 ft

6.
0.5 in 1.3 in 1.2 in

7. A triangle has sides $2\frac{1}{2}$ feet, 3 feet, and $5\frac{1}{4}$ feet

8. A triangle with two equal sides of 3.6 centimeters and the third side 4.1 centimeters

9. A triangle with three equal sides each 5.9 meters

10. A triangle with sides $13\frac{1}{8}$ inches, $11\frac{3}{4}$ inches, and $14\frac{1}{2}$ inches.

Objective 2 Find the area of a triangle.

Find the area of each triangle.

11.
36 m 70 m

12.
15.3 cm 30.4 cm

13.
$7\frac{1}{4}$ ft 6 ft

14.
8 yd 7 yd

15.
5.1 m 6.2 m

16.
$1\frac{3}{8}$ in $\frac{7}{8}$ in $1\frac{1}{4}$ in

Find the shaded area in each figure.

17.

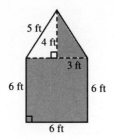

18.

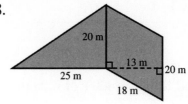

19.

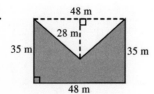

20.

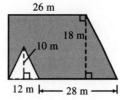

Objective 3 **Given the measure of two angles in a triangle, find the measure of the third angle.**

The measures of two angles of a triangle are given. Find the measure of the third angle.

21. $60°, 70°$

22. $100°, 63°$

23. $30°, 100°$

24. $60°, 60°$

25. $80°, 80°$

26. $37°, 62°$

27. $49°, 72°$

28. $51°, 72°$

29. $77°, 13°$

30. $90°, 45°$

8.5 Mixed Exercises

*Find the **perimeter** of each triangle.*

31.

32.

33.

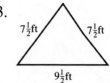

Find the shaded area in each figure.

34.

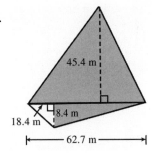

35.

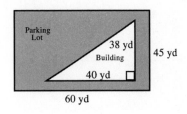

36. If a painter charged \$4.06 per square meter to paint the front of the house shaded below, how much would he charge? All angles that appear to be right angles are indeed right angles.

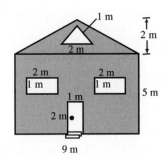

The measures of two angles of a triangle are given. Find the measure of the third angle.

37. $90^o, 50^o$ 38. $41^o, 99^o$ 39. $76^o, 76^o$ 40. $62^o, 57^o$

GEOMETRY

8.6 Circles

Objective 1 **Find the radius and diameter of a circle.**

Find the missing value in each circle.

1.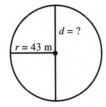

2.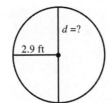

3.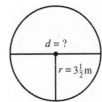

4.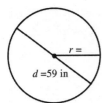

5.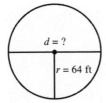

6.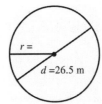

7.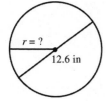

8.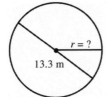

9.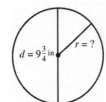

10. The radius of a circle is $\frac{1}{8}$ inch. Find the diameter.

Objective 2 **Find the circumference of a circle.**

Find the circumference of each circle. Use 3.14 as an approximation value for π. Round the answer to the nearest tenth.

11.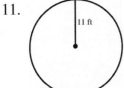

12.

13.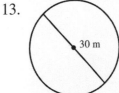

14. A circle with a diameter of 13 inches

15. A circle with a radius of 17 feet

16. A circle with a radius of 4.5 yards

17. A circle with a diameter of 20 centimeters

18. A circle with a diameter of $4\frac{3}{4}$ inches

19. A circle with a radius of $\frac{3}{4}$ mile

20. A circle with a diameter of 12.78 centimeters

Objective 3 **Find the area of a circle.**

Find the area of each circle. Use 3.14 as an approximation value for π. Round the answer to the nearest tenth.

21. 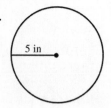   5 in

22. 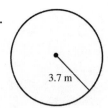   3.7 m

23. 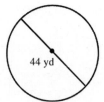   44 yd

24. A circle with diameter of 51 feet

25. A circle with diameter of $5\frac{1}{3}$ yards

26. A circle with diameter of 9.8 centimeters

Find each shaded area. Use 3.14 as an approximation value for π. Round the answer to the nearest tenth if necessary.

27.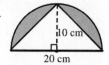
10 cm
20 cm

28.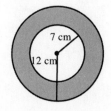
34 m

29.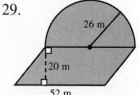
26 m
20 m
52 m

30.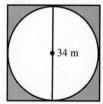
7 cm
12 cm

8.6 Mixed Exercises

Solve each problem.

31. The diameter of a circle is 8 feet. Find its radius.

32. The radius of a circle is 2.7 centimeters. Find its diameter.

33. The diameter of a circle is $12\frac{1}{2}$ yards. Find its radius

Solve each application problem. Use 3.14 as an approximate value for π. Round the answer to the nearest tenth.

34. How far does a point on the tread of a tire move in one turn if the diameter of the tire is 60 centimeters?

35. If you swing a ball held at the end of a string 3 meters long, how far will the ball travel on each turn?

Find the shaded area in each figure. Use 3.14 as an approximation value for π. Round the answer to the nearest tenth.

36.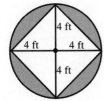

37.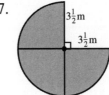

Solve each application problem. Use 3.14 as an approximation value for π. Round the answer to the nearest tenth unless otherwise stated.

38. Find the area of a circular pond that has a diameter of 12.6 meters.

39. Find the cost of sod, at $1.80 per square foot, for the following playing field. Round the answer to the nearest cent.

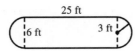

40. Find the area of this skating rink.

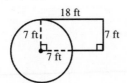

GEOMETRY

8.7 Volume

Objective 1   **Find the volume of a rectangular solid.**

Find the volume of each rectangular solid. Round the answer to the nearest tenth, if necessary.

1.

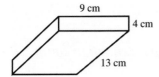

2.

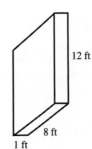

3.

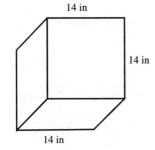

4.

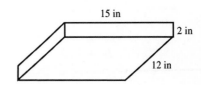

5.

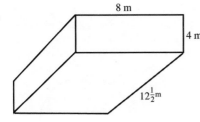

6.

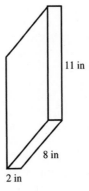

7.

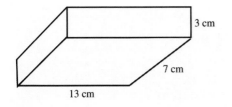

8.

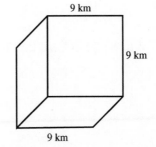

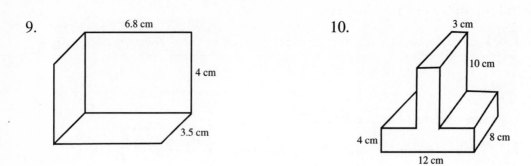

9.   6.8 cm

                              4 cm

                         3.5 cm

10.   3 cm

                        10 cm

           4 cm   8 cm
              12 cm

Objective 2 Find the volume of a sphere.

Find the volume of each sphere or hemisphere. Use 3.14 as an approximation for π. Round the answer to the nearest tenth if necessary.

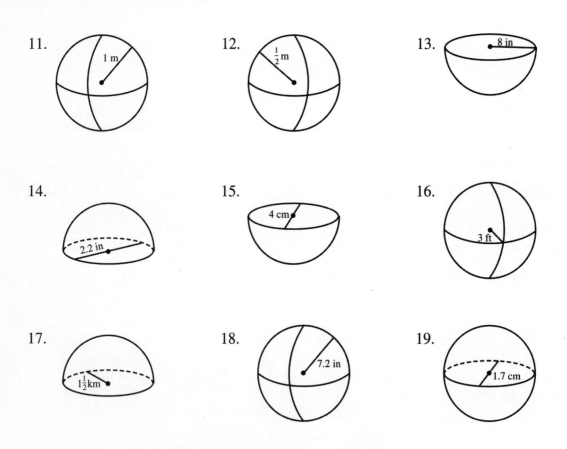

11.   1 m

12. $\frac{1}{2}$ m

13.   8 in

14.   2.2 in

15.   4 cm

16.   3 ft

17. $1\frac{1}{2}$ km

18.   7.2 in

19.   1.7 cm

20. Find the volume of a sphere that has a diameter of $3\frac{1}{4}$ inches.

Objective 3   Find the volume of a cylinder.

Find the volume of each cylinder. Use 3.14 as an approximation value for π. Round the answer to the nearest tenth.

21.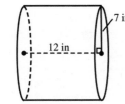

22.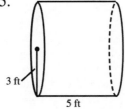

23.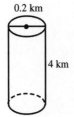

24.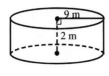

25.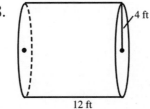

26.

27.

28.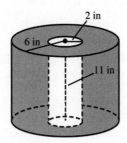

29. Find the volume of the figure. Use 3.14 as an approximation value for π. Round the answer to the nearest tenth.

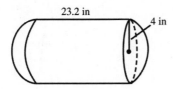

30. Find the volume of the shaded part. Use 3.14 as an approximation value for π. Round the answer to the nearest tenth.

Objective 4 **Find the volume of a cone and a pyramid.**

Find the volume of each figure. Use 3.14 as an approximation value for π. Round the answer to the nearest tenth.

31.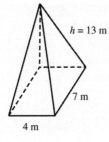

  $h = 13$ m

  7 m

  4 m

32.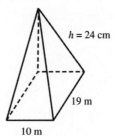

  $h = 24$ cm

  19 m

  10 m

33.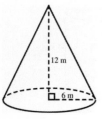

  12 m

  6 m

34.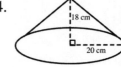

  18 cm

  20 cm

35.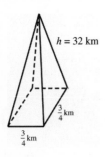

  $h = 32$ km

  $\frac{3}{4}$ km

  $\frac{3}{4}$ km

36.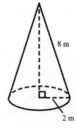

  8 m

  2 m

37.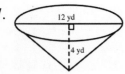

  12 yd

  4 yd

38.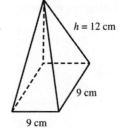

  $h = 12$ cm

  9 cm

  9 cm

39. Find the volume of a pyramid with square base 42 meters on a side and height 38 meters.

40. Find the volume of a cone with base diameter 3.2 centimeters and height 5.8 centimeters.

8.6 Mixed Exercises

Find the volume of each solid which is pictured or described. When necessary use 3.14 as an approximation for π. Round the answer to the nearest tenth, if necessary. Where an area is shaded, find the volume of the shaded part only.

41.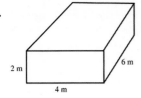

42.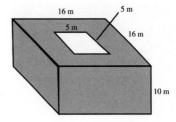

43. A sphere with a diameter of 5 meters

44. A hemisphere with a radius of 11.6 feet

45. A coffee can, radius 6 centimeters and height 16 centimeters

46. A jelly jar, radius 3 centimeters and height 9.1 centimeters

47. An oil can, diameter 8 centimeters and height 13.5 centimeters

48. A cardboard mailing tube, diameter 5 centimeters and height 25 centimeters

49.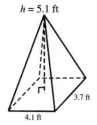

50.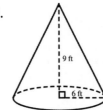

GEOMETRY

8.8 Pythagorean Theorem

Objective 1 Find the square roots using the square root key on the calculator.

Find each square root. Use a calculator with a square root key. Round the answer to the nearest thousandth, if necessary.

1. $\sqrt{17}$ 2. $\sqrt{27}$ 3. $\sqrt{62}$ 4. $\sqrt{55}$

5. $\sqrt{24}$ 6. $\sqrt{10}$ 7. $\sqrt{2}$ 8. $\sqrt{37}$

9. $\sqrt{13}$ 10. $\sqrt{28}$ 11. $\sqrt{47}$ 12. $\sqrt{53}$

13. $\sqrt{71}$ 14. $\sqrt{75}$ 15. $\sqrt{102}$ 16. $\sqrt{145}$

Objective 2 Find the unknown length in a right triangle.

Find the unknown length in each right triangle. Use a calculator with a square root key. Round the answer to the nearest thousandth, if necessary.

17.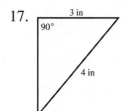

18.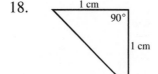

19.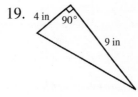

20.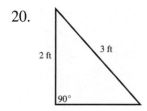

21.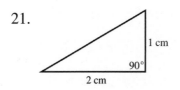

22.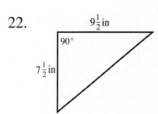

23.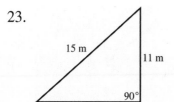

24.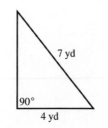

25.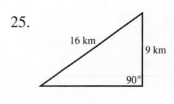

26.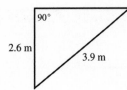

27.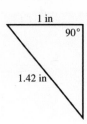

28.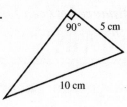

Objective 3 Solve application problems involving right triangles.

Solve each application problem. Use a calculator with a square root key. Round the answer to the nearest thousandth, if necessary.

29. Find the length of this loading dock.

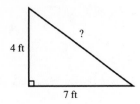

30. Find the unknown length in this roof plan.

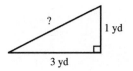

31. A boat goes 10 miles south and then 15

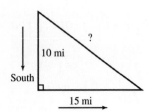

32. Find the height of this telephone pole.

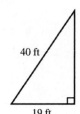

Find the unknown lengths if ladders are leaning against a building as shown. Round the answer to the nearest thousandth, if necessary.

33.

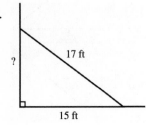

34.

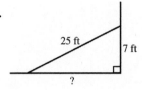

35.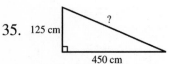

36.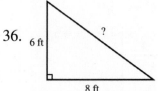

Find the distance between the centers of each pair of holes in each metal plate. Round the answer to the nearest thousandth, if necessary.

37.

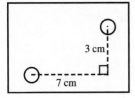

38.

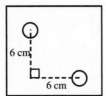

8.8 Mixed Exercises

Find each square root. Use a calculator with a square root key. Round the answer to the nearest thousandth, if necessary.

39. $\sqrt{190}$ 40. $\sqrt{250}$ 41. $\sqrt{153}$ 42. $\sqrt{185}$

Find the unknown length in each triangle. Round the answer to the nearest thousandth, if necessary.

43.

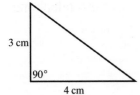

44.

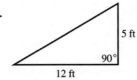

45.

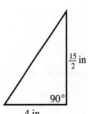

46.

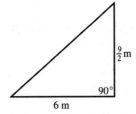

47.

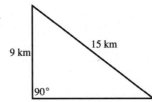

48.

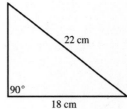

49.

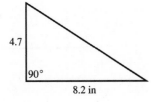

50.

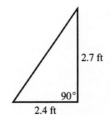

51.

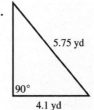

Find the distance between the centers of each pair of holes in each metal plate. Round the answer to the nearest thousandth, if necessary.

52.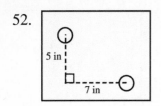

53.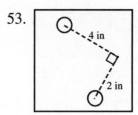

GEOMETRY

8.9 Similar Triangles

Objective 1 Identify corresponding parts in similar triangles.

Name the corresponding angles and the corresponding sides in each pair of similar triangles.

1.
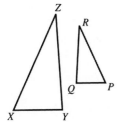

2.
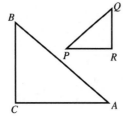

3.
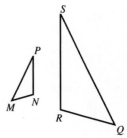

4.
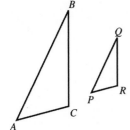

5.
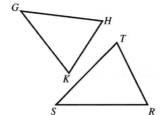

6.
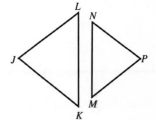

7.
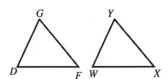

Objective 2 Find the unknown lengths of sides in similar triangles.

Find the unknown length(s) in each pair of similar triangles. Round the answer(s) to the nearest tenth, if necessary.

8.
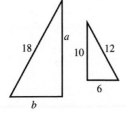

9.
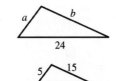

10.
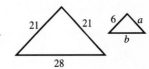

11.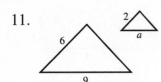

12.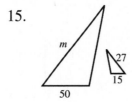

13.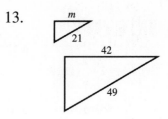

14.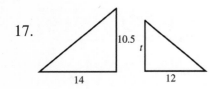

15.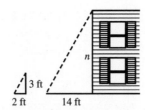

16.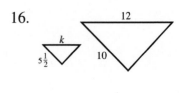

17.

Objective 3   Solve application problems involving similar triangles.

Solve each application problem.

18. The height of the house shown here can be found by using similar triangles and proportion. Find the height of the house by writing a proportion and solving it.

19. A sailor on the US Ramapo saw one of the highest waves ever recorded. He used the height of the ship's mast, the length of the deck and similar triangles to find the height of the wave. Using the information in the figure, write a proportion and then find the height of the wave.

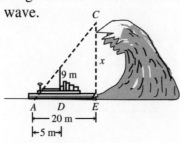

20. Use similar triangles and a proportion to find the length of the lake shown here. (Hint: the side 100 yards long in the smaller triangle corresponds to a side of 100 + 120 = 220 yards in the larger triangle.)

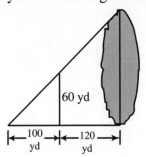

21. To find the height of the tree in the figure, find *y* and then add 1.8 meters for the distance from the ground to eye level.

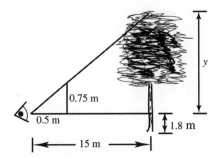

22. One triangle has sides of length 16 meters, 18 meters, and 24 meters. A second triangle, similar to the first, has a shortest side of length 40 meters. Find the lengths of the other two sides of the second triangle.

23. One triangle has sides of length 7.2 centimeters, 9.6 centimeters, and 12 centimeters. A second triangle, similar to the first, has a shortest side of length 4.8 centimeters. Find the lengths of the other two sides of the second triangle.

Find the unknown length in each figure. Round the answer to the nearest tenth, if necessary. Note: When a line is drawn parallel to one side of a triangle, the smaller triangle that is formed will be similar to the original triangle.

24.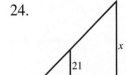

25.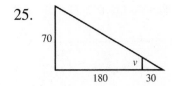

26.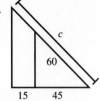

27.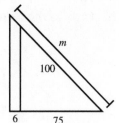

8.9 Mixed Exercises

Each pair of triangles shown is a pair of similar triangles. Name the corresponding angles and corresponding sides.

28.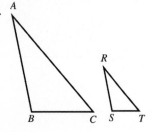

29.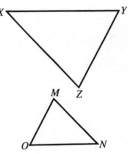

30.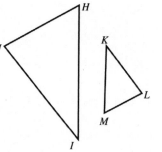

Find the unknown length in each pair of similar triangles. Round the answer to the nearest tenth, if necessary.

31.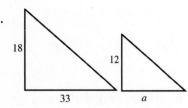

32.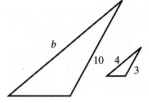

33.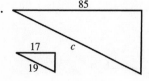

34.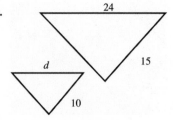

Find the unknown length in each of the following. Round the answer to the nearest tenth, if necessary. Note: When a line is drawn parallel to one side of a triangle, the smaller triangle that is formed will be similar to the original triangle.

35.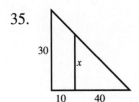

36.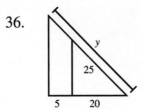

37.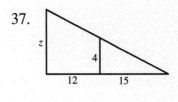

Chapter 9

BASIC ALGEBRA

9.1 Signed Numbers

Objective 1   **Write negative numbers.**

Write a signed number for each of the following.

1. The temperature was 17 degrees above zero.

2. The stock market fell 120 points.

3. The company had a profit of $830.

4. The temperature was 15 degrees below zero.

5. The price fell $13.

6. The rent was increased by $65.

7. The lake was 120 feet below sea level.

8. The mountain was 14,000 feet above sea level.

9. The airplane is 32,000 feet above sea level.

10. The submarine is 185 feet below sea level.

Write **positive, negative,** *or* **neither** *for each of the following.*

11. 16	12. 0	13. −4	14. −6
15. −1.7	16. 3.2	17. $\dfrac{3}{5}$	18. $-\dfrac{7}{5}$
19. $-3\frac{1}{4}$	20. $7\frac{2}{3}$		

Objective 2   **Graph signed numbers on a number line.**

Graph each of the following lists of numbers.

21. −1, 2, 3, 0, −2, 4   22. −4, −2, 3, 5, 0

23. $-\dfrac{1}{2}, -3, -5, \dfrac{1}{4}, 1\frac{7}{8}, 3$

24. $-4, -\dfrac{3}{4}, -2, 4, 1, 3$

25. $-3, -5, -1\frac{1}{2}, \dfrac{2}{3}, 0, 4$

26. $-5, -4, -1, -3\frac{2}{3}, -1\frac{1}{2}, 0, 2$

27. $2, 3\frac{1}{2}, \dfrac{1}{7}, 3, -2\frac{1}{6}$

28. $1, 3\frac{7}{9}, 1\frac{1}{4}, -3\frac{1}{8}, -2$

29. $-12, -14, -13, -9\frac{1}{2}, -6\frac{4}{9}$

30. $-12\frac{2}{3}, -14\frac{1}{5}, -8\frac{3}{7}, -9, -11, -6$

Objective 3   Use the < and > symbols.

Write < or > in each blank to make a true statement.

31. 9 ____ 12

32. 6 ____ 10

33. 12 ____ 5

34. 7 ____ 3

35. −4 ____ 3

36. −8 ____ 7

37. −8 ____ −4

38. 5 ____ −7

39. −11 ____ −5

40. −5 ____ −3

41. −8 ____ −7

42. −23 ____ −32

Objective 4   Find absolute value.

Find the absolute value of the following.

43. $|6|$

44. $|-11|$

45. $-|-8|$

46. $-|352|$

47. $|0|$

48. $\left|-\dfrac{10}{7}\right|$

49. $-|-8.23|$

50. $|5.6|$

51. $|7.2|$

52. $|-8.7|$

53. $-\left|\dfrac{1}{3}\right|$

54. $\left|\dfrac{8}{9}\right|$

Objective 5   Find the opposite of a number.

Find the opposite of each number.

55. 3

56. 8

57. −3

58. −10

59. 12

60. −72

61. −43

62. 41

63. $\dfrac{5}{9}$

64. $\dfrac{9}{10}$

65. $-\dfrac{2}{3}$

66. $-\dfrac{7}{10}$

9.1 Mixed Exercises

Write a singed number for each of the following.

67. The football team lost 7 yards on the first play.

68. The diver was 30 feet below sea level.

69. The price increased $12.

*Write **positive, negative,** or **neither** for each of the following numbers.*

70. –18.76

71. 2.73

Graph each of the following lists of numbers.

72. –2, –1, 0, 1, 2, 5

73. –4.5, –1.5, –0.5, 0, 1.5, 2.5

Write < or > in each blank to make a true statement.

74. 2 ____ –8

75. –3 ____ –9

76. 0 ____ –6

Find the absolute value of the following.

77. $\left| -\dfrac{5}{7} \right|$

78. $\left| -\dfrac{11}{3} \right|$

79. $-\left| -7 \right|$

80. $-\left| -13 \right|$

81. $-\left| 3 \right|$

82. $-\left| 26 \right|$

Find the opposite of each number.

83. 6.1

84. 4.8

85. –2.5

BASIC ALGEBRA

9.2 Adding and Subtracting Signed Numbers

Objective 1 Add signed numbers by using a number line.

Add by using the number line.

1. $4+2$

2. $-1+3$

3. $-5+7$

4. $4+(-7)$

5. $-6+(-1)$

6. $-8+5$

7. $2+(-3)$

8. $-2+(-5)$

9. $-3+(-6)$

10. $2+(-8)$

Objective 2 Add signed number without using a number line.

Add.

11. $-10+17$

12. $-12+7$

13. $-9+(-3)$

14. $-3.2+(-4.7)$

15. $-9.8+(-7.1)$

16. $-7.12+4.3$

17. $-6.21+7.04$

18. $-\dfrac{1}{2}+\dfrac{5}{4}$

19. $\dfrac{7}{6}+\left(-\dfrac{2}{3}\right)$

20. $-\dfrac{7}{5}+\left(-\dfrac{7}{10}\right)$

21. $-4\frac{3}{4}+1\frac{1}{2}$

22. $3\frac{2}{9}+\left(-2\frac{1}{3}\right)$

Objective 3 **Find the additive inverse of a number.**

Give the additive inverse of each number.

23. 4   24. 5   25. −10   26. −15

27. −8   28. −13   29. 77   30. 281

31. 0   32. 3.5   33. 2.7   34. −4.6

Objective 4 **Subtract signed numbers.**

Subtract.

35. $4-11$ 36. $16-19$ 37. $40-95$ 38. $0-11$

39. $0-(-8)$ 40. $-12-0$ 41. $-72-0$ 42. $2.5-3.7$

43. $4.9-8.3$ 44. $4-(-8)$ 45. $-5-(-2)$ 46. $-2-(-5)$

47. $-1-(-5)$ 48. $\dfrac{2}{3}-\left(-\dfrac{7}{12}\right)$ 49. $-\dfrac{1}{2}-\left(-\dfrac{3}{8}\right)$

Objective 5 **Add or subtract a series of signed numbers.**

Follow the order of operations to work each problem.

50. $-3+(-11)-(-2)$ 51. $-8+(-5)-(-3)$ 52. $-6-(-1)+(-9)$

53. $-2-(-9)-(-6)$ 54. $5-(-3)+17$ 55. $4-(-9)-7$

56. $\dfrac{1}{2}-\dfrac{2}{3}+\left(-\dfrac{1}{6}\right)$ 57. $\dfrac{3}{2}+\left(-\dfrac{1}{3}\right)-\left(-\dfrac{5}{6}\right)$ 58. $-3.7-2.5+(-5.3)$

59. $-4.8-(-3.6)+6.4$

9.2 Mixed Exercises

Add by using the number line.

60. $-4+(-2)$ 61. $5+(-6)$

Add.

62. $-12+3$

63. $-61+(-8)$

64. $-9+(-10)$

65. $-8+14$

66. $-23\frac{1}{2}+15$

67. $-5\frac{1}{10}+\left(-2\frac{2}{5}\right)$

Give the additive inverse of each number.

68. -5.11

69. $\dfrac{3}{8}$

70. $\dfrac{15}{7}$

71. $-\dfrac{7}{9}$

72. $-\dfrac{3}{5}$

73. $\dfrac{4}{11}$

Subtract.

74. $-7-9$

75. $-5-12$

76. $-9-(-6)$

77. $-16.2-1.9$

78. $-7.4-5.3$

79. $-12-(-4)$

Follow the order of operations to work each problem.

80. $-1-(-7)-(-6)$

81. $-7-(-9)+(-3)$

82. $5.9-(-4.7)-7.5$

83. $6.8+(-5.9)-(-8.6)$

BASIC ALGEBRA

9.3 Multiplying and Dividing Signed Numbers

Objective 1 **Multiply or divide two numbers with opposite signs.**

Multiply.

1. $-3 \cdot 7$

2. $-6 \cdot 4$

3. $-5 \cdot 10$

4. $9 \cdot (-6)$

5. $4 \cdot (-3)$

6. $8 \cdot (-12)$

7. $-\dfrac{11}{15} \cdot \dfrac{25}{22}$

8. $-\dfrac{5}{9} \cdot \dfrac{27}{20}$

9. $9 \cdot (-3.7)$

10. $14 \cdot (-2.3)$

11. $-7.1 \cdot (3.8)$

12. $-6.2 \cdot (4.8)$

Divide.

13. $\dfrac{-16}{8}$

14. $\dfrac{28}{-7}$

15. $\dfrac{20}{-5}$

16. $-96 \div 12$

17. $-\dfrac{5}{6} \div \dfrac{2}{3}$

18. $5 \div \left(-\dfrac{5}{11}\right)$

Objective 2 **Multiply or divide two numbers with the same sign.**

Multiply.

19. $-11 \cdot (-4)$

20. $-15 \cdot (-6)$

21. $-21 \cdot (-2)$

22. $-18 \cdot (-1)$

23. $-\dfrac{1}{4} \cdot (-10)$

24. $-\dfrac{2}{5} \cdot (-6)$

Divide.

25. $\dfrac{-65}{-5}$

26. $\dfrac{-54}{-27}$

27. $\dfrac{-72}{-9}$

28. $\dfrac{-35}{-5}$

29. $-\dfrac{5}{8} \div -\dfrac{10}{3}$

30. $-\dfrac{7}{18} \div -\dfrac{7}{9}$

31. $-\dfrac{2}{3} \div -2$

32. $\dfrac{-\frac{3}{11}}{-\frac{7}{33}}$

33. $\dfrac{-\frac{2}{7}}{-\frac{8}{21}}$

9.3 Mixed Exercises

Multiply or divide as indicated.

34. $4 \cdot (-20)$
35. $13 \cdot (-3)$
36. $9 \cdot (-12)$
37. $7 \div \left(-\dfrac{21}{8} \right)$

38. $\dfrac{-11.84}{2}$
39. $\dfrac{-59.2}{7.4}$
40. $-\dfrac{6}{5} \cdot \left(-\dfrac{15}{4} \right)$
41. $-\dfrac{7}{10} \cdot \left(-\dfrac{5}{4} \right)$

42. $-15 \cdot \left(-\dfrac{3}{10} \right)$
43. $\dfrac{-\frac{9}{20}}{-\frac{3}{4}}$
44. $-\dfrac{5}{8} \div (-10)$
45. $\dfrac{-14.88}{-3.1}$

46. $\dfrac{-22.75}{-5}$

BASIC ALGEBRA

9.4 Order of Operations

Objective 1 Use the order of operations.

Simplify each of the following.

1. $-5+6+(-3)\cdot 7$

2. $4+(-3)+2\cdot(-5)$

3. $-3-10\div 2$

4. $6-12\div 3$

5. $8+3\cdot(-6)$

6. $6-2\cdot(-11)$

7. $(4-7)\cdot(-2)$

8. $(3-9)\cdot(-5)$

Objective 2 Use the order of operations with exponents.

Simplify each of the following.

9. $(-2)^2+6^2$

10. $10^2+(-5)^2$

11. $36\div(-3)\div 2^2$

12. $-54\div(-2)\div 3^2$

13. $-4\cdot 2^2-5\cdot 3-(-6)$

14. $3\cdot 5^2-3\cdot 7-9$

15. $-\left(\dfrac{3}{7}+\dfrac{2}{7}\right)+\left(-\dfrac{1}{3}\right)^2$

16. $-\left(\dfrac{2}{3}-\dfrac{1}{3}\right)+\left(\dfrac{1}{2}\right)^2$

17. $-\left(-\dfrac{1}{6}+\dfrac{5}{6}\right)\div\left(-\dfrac{1}{3}\right)^2$

18. $-\left(-\dfrac{4}{7}\right)-\left(-\dfrac{2}{7}\right)\div\left(\dfrac{1}{7}\right)^2$

Objective 3 User the order of operations with fraction bars.

Simplify each of the following.

19. $\dfrac{3-(-7)-5\cdot 4}{(2-5)^2-(-1)}$

20. $\dfrac{4-(-6)-4\cdot 6}{(3-5)^2-(-3)}$

21. $\dfrac{-9-(-3-4)}{(-3)^2-7}$

22. $\dfrac{-8-(-5-7)}{5-(-3)^2}$

9.4 Mixed Exercises

Simplify each of the following.

23. $(-3+5)\cdot(6+1)$

24. $(11+9)\cdot(-7+3)$

25. $-2\cdot3+4\div2-3$

26. $8\div(-4)+4\cdot(-3)+2$

27. $-3\cdot(8-16)\div(-8)$

28. $-5\cdot(8-14)\div(-10)$

29. $2^2\cdot3^2+(-3)\cdot2+1$

30. $3^3+5\cdot(-3)+3\cdot4$

31. $(-2)^4-4\cdot2-2\cdot5$

32. $2-6^2\div2^2-5$

33. $6^2\div3^2-4\cdot3-2\cdot5$

34. $\left(-\dfrac{1}{2}\right)^2-\left(\dfrac{3}{4}-\dfrac{7}{4}\right)$

35. $\left(-\dfrac{1}{5}\right)^2-\left(\dfrac{4}{5}-\dfrac{1}{5}\right)$

36. $-\dfrac{2}{3}-\dfrac{1}{3}-\left(\dfrac{1}{5}\right)^2$

37. $\dfrac{(-5)^2+(-3-4)}{-3-2\cdot3}$

38. $\dfrac{(-4-7)-(-3)^2}{-4-2\cdot3}$

BASIC ALGEBRA

9.5 Evaluating Expressions and Formulas

Objective 2 Find the value of an expression when values of the variable are given.

Find the value of the expression $3r - 2s$ *for each of the following values of r and s.*

1. $r = 2, s = 5$

2. $r = 1, s = 7$

3. $r = 4, s = -6$

4. $r = -3, s = 4$

5. $r = -7, s = -3$

6. $r = -5, s = -8$

7. $r = 0, s = -12$

8. $r = -4, s = 0$

Use the given values of the variables to find the value of each expression.

9. $-4k + 3m; k = 5, m = -\dfrac{1}{3}$

10. $-2d + f; d = \dfrac{1}{2}, f = -3$

11. $\dfrac{2y - z}{2 - x}; y = 1, z = 6, x = 4$

12. $\dfrac{-a + 3b}{c - 2}; a = 4, b = -1, c = -5$

In each of the following, use the given formula and values of the variables to find the value of the remaining variable.

13. $P = 4s; s = 5$

14. $P = a + b + c; a = 7, b = 8, c = 5$

15. $P = 2L + 2W; L = 8, W = 6$

16. $P = 2L + 2W; L = 15, W = 11$

17. $A = \dfrac{1}{2}bh; b = 8, h = 9$

18. $A = \dfrac{1}{2}bh; b = 6, h = 13$

19. $V = \dfrac{1}{3}Bh; B = 50, h = 3$

20. $V = \dfrac{1}{3}Bh; B = 102, h = 2$

21. $d = rt; r = 65, t = 3$

22. $d = rt; r = 220, t = 4$

23. $C = 2\pi r; \pi \approx 3.14, r = 9$

24. $C = 2\pi r; \pi \approx 3.14, r = 12$

BASIC ALGEBRA

9.6 Solving Equations

Objective 1 Determine whether a number is a solution of an equation.

Decide whether the given number is a solution of the equation.

1. $p+2=4$; 2

2. $a-3=1$; 4

3. $b-5=18$; 13

4. $4x=12$; 4

5. $6y=42$; 8

6. $-3c=-12$; 4

7. $-11c=33$; 3

8. $\frac{1}{2}k=-4$; -8

9. $\frac{1}{3}d=12$; 4

10. $2-3m=2$; 0

11. $-x=-3$; 3

12. $y=-3$; 3

Objective 2 Solve equations using the addition property of equations.

Solve each equation by using the addition property. Check each solution.

13. $x+6=10$

14. $k-3=7$

15. $p-5=9$

16. $12=x+7$

17. $k+16=27$

18. $y+11=16$

19. $7-y=9$

20. $9=-4+x$

21. $-2+m=-1$

22. $-7+r=-9$

23. $-6=-2+y$

24. $-12=-10+a$

25. $x+\dfrac{1}{3}=1$

26. $z+\dfrac{5}{8}=3$

Objective 3 Solve equations using the multiplication property of equations.

Solve each equation. Check each solution.

27. $15a=60$

28. $77=11m$

29. $3x=-18$

30. $7x=-42$

31. $-6k=-42$

32. $-32=-2p$

33. $\dfrac{k}{2}=16$

34. $-2=\dfrac{x}{7}$

35. $\dfrac{a}{-5}=10$

36. $3=\dfrac{m}{-7}$

37. $-12=\dfrac{r}{3}$

38. $\dfrac{1}{2}p=6$

9.6 Mixed Exercises

Decide whether the given number is a solution of the equation.

39. $5 + 8m = 3$; -1

40. $2z - 3 = -2$; $\dfrac{1}{2}$

41. $-5y + 1 = 6$; -1

42. $-6k - 3 = 7$; $\dfrac{2}{3}$

Solve each equation. Check each solution.

43. $m - \dfrac{3}{4} = 11$

44. $6 = k - \dfrac{2}{3}$

45. $a - 5 = \dfrac{2}{3}$

46. $s - 4 = \dfrac{5}{9}$

47. $x - 1.24 = 4.37$

48. $6.99 = a - 3.27$

49. $4.76 + r = 2.15$

50. $-9x = -63$

51. $1.32 = -1.2m$

52. $-3.2y = -8.32$

53. $-\dfrac{1}{4}m = 8$

54. $-16 = -\dfrac{4}{5}x$

55. $\dfrac{3}{5}a = \dfrac{1}{4}$

56. $\dfrac{1}{4} = \dfrac{2}{3}c$

57. $\dfrac{k}{4.2} = 0.5$

58. $-1.1 = \dfrac{m}{-5.2}$

BASIC ALGEBRA

9.7 Solving Equations with Several Steps

Objective 1 Solve equations with several steps.

Solve each equation. Check each solution.

1. $5p - 4 = 1$

2. $6k + 4 = 16$

3. $13 = 2y - 9$

4. $4x - 3 = 5$

5. $-3m + 2 = -4$

6. $-6 = -4k + 6$

7. $31 = -8a + 7$

8. $-2p + 5 = 13$

9. $-6a - 7 = 11$

10. $33 = -12a - 3$

Objective 2 Use the distributive property.

Use the distributive property to simplify.

11. $8(5 + 2)$

12. $17(10 - 1)$

13. $3(x + 7)$

14. $7(k - 5)$

15. $-3(6 + m)$

16. $-5(2 + a)$

17. $-2(y - 3)$

18. $-1(5 + a)$

19. $-3(5 - a)$

20. $\dfrac{1}{2}(10 - 2x)$

Objective 3 Combine like terms.

Combine like terms.

21. $10r + 6r$

22. $-3m + 5m$

23. $9x + 7x$

24. $13y + 5y$

25. $-32m - 4m$

26. $9z - 4z$

27. $4k - 12k$

28. $15a - 20a$

29. $3z - 10z$

30. $-3.2x + 1.3x$

Objective 4   Solve more difficult equations.

Solve each equation. Check each solution.

31. $8a + 4a = 60$

32. $6x + x = 42$

33. $18m - 11m = 35$

34. $21 = 9z - 6z$

35. $-20 = 3y - 7y$

36. $2z - 9z = -56$

37. $6y - 13y = -14$

38. $11z - 17z = -18$

39. $-3.6m - 8 = 2.4m - 2$

40. $-8.2p + 3 = 1.8p - 7$

41. $\dfrac{1}{2}y - 6 = \dfrac{1}{4}y - 1$

42. $\dfrac{2}{3}z + 2 = \dfrac{1}{2}z - 1$

43. $-3(x - 2) = 12$

44. $-4(x + 3) = 16$

45. $-5(3 - x) = 25$

46. $-7 = 7(3 - y)$

47. $26 = 13(2 - a)$

47. $26 = 13(2 - a)$

48. $\dfrac{1}{2}(b + 3) = -4$

49. $0.6 = 0.3(0.2 - c)$

50. $-1.2(x - 2) = 3.6$

51. $-3 + 5 - 7 = -7a + 5a$

52. $-4.1y - y = -2.3 - 2.8$

9.7 Mixed Exercises

Use the distributive property to simplify.

53. $-5(7 - r)$

54. $-4(8 - x)$

Combine like terms.

55. $5a - 2.7a$

56. $\dfrac{1}{3}k - \dfrac{5}{6}k$

Solve each equation. Check each solution.

57. $-\dfrac{1}{2}z + 1 = -2$

58. $-\dfrac{5}{8}r + 6 = -4$

59. $2p - 3.1 = 6.5$

60. $5p - 4.2 = -17.7$

61. $0.2x + 4.1 = 3.7$

62. $0.3x - 12.5 = -14.6$

63. $6p - 4 = 4p + 10$

64. $7z - 5 = 4z - 14$

65. $7z - 9 = 10z + 9$

66. $3x - 7 = 9x - 13$

67. $2.5y + 7 = 4.5y + 11$

68. $4.2x + 1 = 1.2x - 8$

69. $-3.7z - 0.5 = -5.2z + 4$

70. $-2.1k + 4.7 = -1.7k + 1.5$

71. $-6(3 - 2d) = 10$

72. $-12 = 0.2(5 - x)$

BASIC ALGEBRA

9.8 Using Equations to Solve Application Problems

Objective 1 **Translate word phrases into expressions with variables.**

Write each word phrase in symbols, using x as the variable.

1. 13 plus a number

2. The sum of 9 and a number

3. 5 added to a number

4. A number increased by –9

5. 5 subtracted from a number

6. 6 fewer than a number

7. The product of a number and 3

8. Double a number

9. Twice a number added to 7

10. The sum of three times a number and 3

11. Three times a number added to five times the number

12. Six times a number subtracted from ten times the number.

Objective 2 **Translate sentences into equations.**

Translate each sentence into an equation and solve it. Check your solution by going back to the words in the original problem.

13. If –5 times a number is added to 4, the result is –11. Find the number.

14. If twice a number is subtracted from 45, the result is 35. Find the number.

15. If three times a number is added to 7, the result is 1. Find the number.

16. The sum of 6 and four times a number is 50. Find the number.

17. When twice a number is decreased by 3, the result is –17. Find the number.

18. If the product of some number and 2 is increased by 18, the result is four times the number. Find the number.

Objective 3 Solve application problems

Solve each application problem. Show your work for each of the six problem-solving steps.

19. If four times a number is decreased by 3, the result is 13. Find the number.

20. When 9 is added to twice a number, the result is 15. Find the number.

21. If twice a number is added to four times the number, the result is 42. Find the number.

22. When three times a number is subtracted from seven times the number, the result is 20. Find the number.

23. If half a number is added to twice the number, the result is 55. Find the number.

24. If one third of a number is added to three times the number, the result is 50. Find the number.

25. Twice a number is added to the number giving –30. Find the number.

26. If ten times a number is subtracted from six times the number, the result is 12. Find the number.

27. A board is 91 centimeters long. It is to be cut into two pieces, with one piece 15 centimeters longer than the other. Find the length of the shorter piece.

28. Lisa and Michael were opposing candidates for city council. Lisa won, with 73 more votes than Michael. The total number of votes received by both candidates was 567. Find the number of votes received by Michael.

29. A chain saw can be rented for $7 a day plus a one-time $8 sharpening fee. The bill for a rental was $43. For how many days was the saw rented?

9.8 Mixed Exercises

Write each word phrase in symbols, using x as the variable.

30. The sum of a number and 7

31. The total of a number and 3

32. 8 less than a number

33. A number subtracted from 1

34. Triple a number

35. Half a number

36. A number divided by 4

37. 7 divided by a number

Translate each sentence into an equation and solve it. Check your solution by going back to the words in the original problem.

38. If three times a number is decreased by 12, the result is 3. Find the number.

39. If seven times a number is subtracted from nine times a number, the result is 16. Find the number.

40. When two times a number is subtracted from 8, the result is 20 plus the number. Find the number.

41. If half a number is added to 4, the result is 10. Find the number.

Solve each application problem. Show your work for each of the six problem-solving steps.

42. A rental car costs $32 per day plus $0.20 per mile. The bill for a one-day rental was $82. How many miles was the care driven?

43. The perimeter of a rectangle is 46 meters. The width is 9 meters. Find the length.

44. The length of a rectangle is 27 centimeters, while the perimeter is 70 centimeters. Find the width of the rectangle.

45. A fence is 694 meters long. It is to be cut into three parts. Two of the parts are the same length, while the third part is 25 meters longer than the other two. Find the length of each of the equal parts.

46. A wooden railing is 87 meters long. It is to be divided into four pieces. Three of the pieces will be the same length, and the fourth piece will be 3 meters longer than each of the other three. Find the length that each of the three pieces of equal length will be.

47. For how long must $700 be deposited at 12% per year to earn $420 interest?

48. How much money must be deposited at 13% per year for 3 years to earn $2730 interest?

Chapter 10

STATISTICS

10.1 Circle Graphs

Objective 1 Read and understand a circle graph.

The circle graph shows the enrollment by major at a small college.

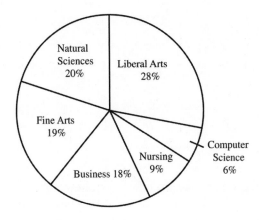

The total enrollment at the college is 3200 students. Use the circle graph to find the number of students with each of the following majors.

1. Liberal Arts

2. Natural Sciences

3. Fine Arts

4. Business

5. Nursing

6. Computer Science

7. What is the most popular major at the college?

8. What major has the fewest students?

9. Find the ratio of business majors to computer science majors.

10. Find the ratio of natural science majors to liberal arts majors.

Objective 2   **Use a circle graph.**

The circle graph shows the costs of remodeling a kitchen. Use the graph to solve Problems 11–16.

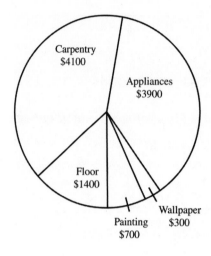

11. Find the total cost of remodeling the kitchen.

12. What is the largest single expense in remodeling kitchen?

13. Find the ratio of the cost of appliances to the total remodeling cost.

14. Find the ratio of the cost of painting to the total remodeling cost.

15. Find the ratio of the cost of wallpaper to the cost of the floor.

16. Find the ratio of the cost of painting to the cost of the floor.

The following circle graph shows the number of students at a college enrolled in certain majors.
Use the graph to solve Problems 17–22.

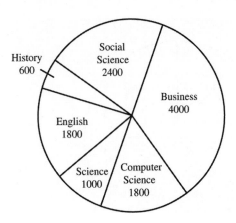

17. Which major has the most number of students enrolled?

18. Find the ratio of the number of business majors to the total number of students.

19. Find the ratio of the number of English majors to the total number of students.

20. Find the ratio of the number of science majors to the number of English majors.

21. Find the ratio of the number history majors to the number of social science majors.

22. Find the ratio of the number of computer science majors to the number of business majors.

The circle graph shows the expenses involved in keeping a sales force on the road. Each expense item is expressed as a percent of the total sales force cost of $950,000. Find the number of dollars of expense for each category in Problems 23–28.

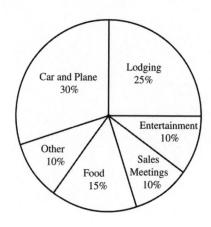

23. Car and plane

24. Lodging

25. Entertainment

26. Sales meetings

27. Food

28. Other

Objective 3 **Draw a circle graph.**

During one semester, Evie Allsot, a student spent $1600 for expenses as shown in the following chart. Complete the chart.

Item	Dollar Amount	Percent of Total	Degrees of a Circle
Food	$400	25%	90°
29. Rent	$320	_____	72°
30. Clothing	$240	_____	_____
31. Books	$160	10%	_____
32. Entertainment	$240	_____	54°
33. Savings	$80	_____	_____
34. Other	_____	_____	36°

35. Draw a circle graph using the above information.

10.1 Mixed Exercises

The circle graph shows how the Recycling Club's income of $42,800 is budgeted. Find the number of dollars budgeted for each category in Problems 36–41.

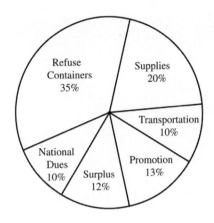

36. Supplies

37. Refuse containers

38. Promotion

39. Surplus

40. National dues

41. Transportation

42. Jensen Manufacturing Company has its annual sales divided into five categories as follows.

Item	Annual Sales
Parts	$20,000
Hand tools	80,000
Bench tools	100,000
Brass fittings	140,000
Cabinet hardware	60,000

(a) Find the total sales for a year.

(b) Find the percent of the total sales for each item.

(c) Find the number of degrees in a circle graph for each item.

(d) Make a circle graph showing this information.

43. A book publisher had 30% of its sales in mysteries, 15% in biographies, 10% in cookbooks, 25% in romantic novels, 15% in science, and the rest in business books.

(a) Find the number of degrees in a circle graph for each type of book.

(b) Draw a circle graph.

44. A family recorded its expenses for a year, with the following results.

Item	Percent of Total
Housing	40%
Food	20%
Automobile	14%
Clothing	8%
Medical	6%
Savings	8%
Other	4%

(a) Find the number of degrees in a circle graph for each item.

(b) Draw a circle graph.

STATISTICS

10.2 Bar Graphs and Line Graphs

Objective 1 **Read and understand a bar graph.**

The bar graph sows the enrolment for the fall semester at a small college for the past five year.

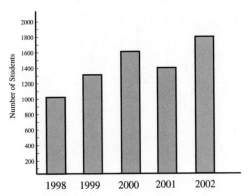

Use the bar graph above for Problems 1–10. Find the enrollment for the fall semester for the following years.

1. 1998   2. 1999   3. 2000   4. 2001   5. 2002

6. How many more students were enrolled in 2000 than in 1999?

7. What year had the greatest enrollment?

8. Which year showed a decrease in enrollment?

9. By how many students did the enrollment increase from 1998 to 1999?

10. Which year showed the greatest increase in enrollment?

Objective 2 **Read and understand a double bar graph.**

The double-bar graph shows the enrollment by gender in each class at a small college. Use the double-bar graph for Problems 11–20.

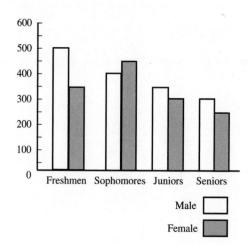

11. Which class has a greater female enrollment than male enrollment?

12. How many female freshmen are enrolled?

13. How many male sophomores are enrolled?

14. Find the total number of juniors enrolled.

15. Find the total number of freshmen enrolled.

16. Find the ratio of freshmen males to freshmen females.

17. Find the ratio of sophomore females to senior females.

18. Find the total number of students enrolled.

19. Find the ratio of freshmen students to senior students.

20. Which class has the greatest difference between male students and female students?

Objective 3 **Read and understand a bar graph.**

The line graph gives the value of one share of stock of Microchip Computer Corporation on the first trading day of the month for six consecutive months. Use the line graph for Problems 21–30.

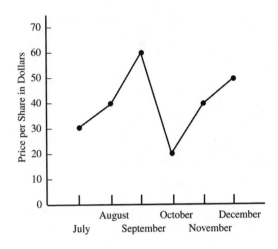

21. In which month was the value of the stock highest?

22. In which month was the value of the stock lowest?

23. Find the value of one share on the first trading day October.

24. Find the value of one share on the first day of trading of July.

25. Find the increase in the value of one share from October to November.

26. What is the largest monthly decrease in the value of one share.

27. Find the ratio of the value of one share on the first trading day in September to the value of one share on the first trading day of October.

28. Comparing the value of one share on the first trading day in July to the first trading day in November, has the value increased, decreased, or remained unchanged?

29. By how much did the value of one share increase from July to September?

30. By how much did the value of one share increase from October to December?

Objective 4 **Read and understand a comparison line graph.**

The comparison line graph shows annual sales for two different stores for each of the past few years. Use the graph to solve Problems 31–40.

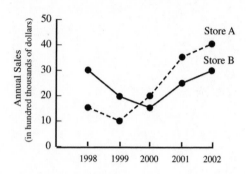

Find the annual sales for store A in each of the following years.

31. 2001   32. 1999   33. 1998

Find the annual sales for store B in each of the following years.

34. 2001   35. 1999   36. 1998

37. In which years did the sales of store A exceed the sales of store B?

38. Which year showed the least difference between the sales of store A and the sales of store B?

39. Which year showed the greatest difference between the sales of store A and the sales of store B?

40. Find the ratio of the sales of store A to the sales of store B in 1998.

10.2   Mixed Exercises

The bar graph shows the attendance of the County Fair for five days in July. Use the graph to solve Problems 41–50.

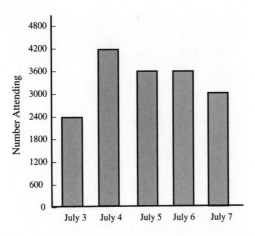

41. Find the attendance on July 5.

42. Which day had the lowest attendance?

43. Find the attendance on July 3.

44. Find the attendance on July4.

45. Which day had the greatest attendance?

46. Find the attendance on July 7.

47. How many more people attended on July 4 than on July 3?

48. How many more people attended on July 6 than on July 7?

49. Which two days had the same attendance?

50. What was the total attendance for the five days?

The double-bar graph shows the number of employees at one company during the first six months of 2001 and 2002. Use the graph to solve Problems 51–56.

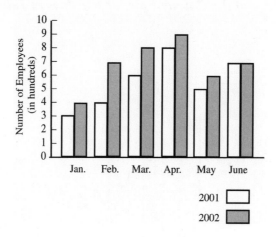

51. Which month in 2002 had the greatest number of people employed? What was the total in that month?

52. How many employees were there in January of 2001?

53. How many more employees were there in February of 2002 than in February of 2001?

54. How many more employees were there in March of 2002 than in March of 2001?

55. Find the increase in the number of employees from February 2001 to April 2001.

56. Find the increase in the number of employees from January 2002 to May 2002.

The double-bar graph shows sales of unleaded and supreme unleaded gasoline at a service station for a five-year period. Use the graph to solve Problems 57–62.

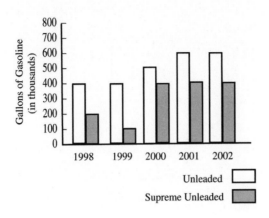

57. How many gallons of unleaded gasoline were sold in 2001?

58. How many gallons of supreme unleaded gasoline were sold in 1998?

59. The greatest difference in sales between supreme unleaded and unleaded gasoline was in which year?

60. The smallest difference in sales between supreme unleaded and unleaded gasoline was in which year?

61. Find the increase in unleaded gasoline sales from 1998 to 2002.

62. Find the increase in supreme unleaded gasoline from 1998 to 2002.

The line graph shows the number of burglaries in a community during the first six months of a year. Use the graph to solve Problems 63–72.

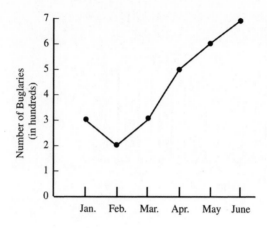

63. Which month had the greatest number of burglaries?

64. Which month had the least number of burglaries?

65. Find the number of burglaries in April.

66. Find the number of burglaries in February.

67. Find the increase in the number of burglaries from March to April.

68. Find the number burglaries in January.

69. Find the decrease in the number of burglaries from January to February.

70. What is the trend in the number of burglaries from January to June? That is, is there an overall increase or decrease in the number of burglaries?

71. How many more or less burglaries were there in June than in January?

72. What is the total number of burglaries in the six months?

The comparison line graph sows sales and profits for a fast food operation for each of the past few years. Use the graph to solve Problems 73–78.

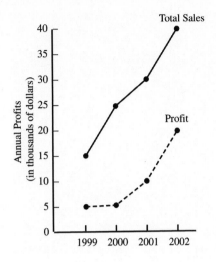

Find the total sales in each of the following years.

73. 1999   74. 2001   75. 2002

Find the profit in each of the following years.

76. 1999   77. 2001   78. 2002

STATISTICS

10.3 Frequency Distributions and Histograms

Objective 1 **Understand a frequency distribution.**

The following scores were earned by students on an algebra exam. Use the data to complete the table.

84	90	83	72	84	93	83	90	83
90	72	64	90	83	72	83	83	64

	Score	Tally	Frequency
1.	64		
2.	72		
3.	83		
4.	84		
5.	90		
6.	93		

Objective 2 **Arrange data in class intervals.**

The following list of numbers represents systolic blood pressure of 21 patients. Use these numbers to complete the table.

120	98	180	128	143	98	105
136	115	190	118	105	180	112
160	110	138	122	98	175	118

	Systolic Blood Pressure	*Tally*	*Frequency*
7.	90–109		
8.	110–129		
9.	130–149		
10.	150–169		
11.	179–189		
12.	190–209		

13. What was the most common range of systolic blood pressure?

14. What was the least common range of systolic blood pressure?

Objective 3 **Read and understand a histogram.**

A local chess club recorded the ages of their members and constructed a histogram. Use the histogram to solve Problems 15–20.

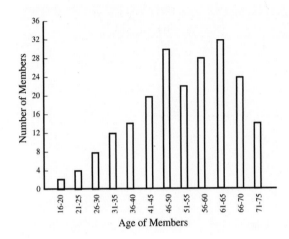

15. The greatest number of members is in which age group?

16. The fewest number of members are in which age group?

17. Find the number of members 30 years of age or younger.

18. Find the number of members 51 years and older.

19. How many members are 51–65 years of age?

20. How many members are 46–50 years of age?

10.3 Mixed Exercises

The following list of numbers represents IQ scores of 18 students. Use these numbers to complete the following table.

98	121	112	99	105	112
110	100	92	109	104	106
105	88	92	103	98	118

	IQ Scores	*Tally*	*Frequency*
21.	80–89		
22.	90–99		
23.	100–109		
24.	110–119		
25.	120–129		

26. What was the most common range of IQ scores?

27. What was the least common range of IQ scores?

28. The following list of numbers represents systolic blood pressures of 21 patients. Construct a histogram from this data. Use intervals 90–109, 110–129, 130–149, 150–169, 170–189, and 190–209.

120	98	180	128	143	98	105
136	115	190	118	105	180	102
160	110	138	122	98	175	118

29. The following list of numbers represents IQ scores of 18 students. Construct a histogram from this data. Use intervals 80–89, 90–99, 100–109, 110–119, and 120–129.

98	121	112	99	105	112
110	100	92	109	104	106
105	88	92	103	98	118

STATISTICS

10.4 Mean, Median, and Mode

Objective 1 Find the mean of a list of numbers.

Find the mean for each list of numbers. Round the answer to the nearest tenth, if necessary.

1.   7, 12, 3, 5, 9

2.   51, 47, 33, 43, 79, 58

3.   39, 50, 59, 61, 69, 73, 51, 80

4.   31, 37, 44, 51, 52, 74, 69, 83

5.   3.8, 9.2, 6.7, 3.5, 4.9, 8.8

6.   62.7, 59.6, 71.2, 65.8, 63.1

7.   19900, 23850, 25930, 27710, 29340, 41000

8.   48500, 49720, 32183, 19982, 52250

9.   8.5, 11.3, 9.5, 8.3, 9.2, 8.4, 9.7, 4.3, 3.2, 4.7

10. 40.1, 32.8, 82.5, 51.2, 88.3, 31.7, 43.7, 51.2

Objective 2 Find a weighted mean.

Find the weighted mean. Round the answer to the nearest tenth, if necessary.

11.

Value	Frequency
2	4
6	2
9	1
13	3

12.

Value	Frequency
17	4
12	5
15	3
19	1

13.

Value	Frequency
13	4
12	2
19	5
15	3
21	1
27	5

14.

Value	Frequency
35	1
36	2
39	5
40	4
42	3
43	5

15.

Value	Frequency
1	2
2	3
4	5
5	7
6	4
7	2
8	1
9	1

16.

Value	Frequency
21	2
23	1
25	3
26	4
28	5
30	3
31	2
33	1

Find the grade point average for each of the following students. Assume A = 4, B = 3, C = 2, D = 1, F = 0. Round to the nearest tenth, if necessary.

17.

Units	Grade
4	C
2	B
5	C
1	D
3	F

18.

Units	Grade
3	C
3	A
4	B
5	B
2	A

19.

Units	Grade
5	B
4	C
3	B
2	C
2	C

20.

Units	Grade
3	A
4	B
2	C
5	C
2	D

Objective 3 **Find the median.**

Find the median for each list of numbers.

21. 13, 19, 33, 52, 93, 107

22. 199, 472, 312, 298, 254

23. 200, 215, 226, 238, 250, 283

24. 30.0, 28.2, 28.8, 32.6

25. .002, .004, .012, .008

26. 1.6, 1.8, 1.7, 1.1, 1.7

27. 389, 464, 521, 610, 654, 672, 682, 712

28. 43, 69, 108, 32, 51, 49, 83, 57, 64

29. 2073, 2069, 2095, 2053, 2067, 2016, 2019

30. 948, 923, 998, 912, 984, 926

Objective 4 **Find the mode.**

31. 3, 8, 7, 5, 8, 1, 6, 2

32. 32, 43, 57, 43, 59, 43, 57

33. 5, 4, 5, 1, 3, 6, 9, 7, 5, 2

34. 4, 9, 3, 4, 7, 3, 2, 3, 9

35. 85, 79, 79, 79, 86, 86, 85, 85, 81

36. 238, 272, 274, 272, 268, 271

37. 4, 8, 16, 2, 1, 7, 18, 9, 3, 19

38. 13, 16, 18, 19, 22, 30, 33, 85, 90

39. 37, 24, 35, 35, 24, 38, 39, 28, 27, 39

40. 172.6, 199.7, 182.4, 167.1, 172.6, 183.4, 187.6

10.4 Mixed Exercises

Find the mean for each list of numbers. Round the answer to the nearest tenth, if necessary.

41. 12, 13, 7, 11, 19

42. 2.8, 3.9, 4.7, 5.6, 6.5, 9.1

43. 216, 245, 268, 268, 280, 291, 304, 313

Find the weighted mean. Round the answer to the nearest tenth, if necessary.

44.

Value	Frequency
12	2
14	5
16	7
18	6
20	4

45.

Value	Frequency
30	2
31	5
32	7
33	4
34	3
35	1

Find the median for each list of numbers.

46. 21, 32, 27, 23, 25, 29, 22

47. 1.8, 1.2, 1.1, 1.9, 2.6

48. 200, 195, 302, 284, 256, 237, 239, 240

Find the mode or modes for each list of numbers.

49. 2, 4, 6, 6, 8, 10, 8, 12, 14, 8

50. 0.2, 0.7, 0.9, 0.7, 0.5, 0.3, 0.4, 0.7, 0.2

ANSWERS TO
ADDITIONAL EXERCISES

Chapter 1

WHOLE NUMBERS

1.1 Reading and Writing Whole Numbers

Objective 1

1. Whole number

2. Not a whole number

3. Not a whole number

4. Whole number

5. Whole number

6. Whole number

7. Not a whole number

8. Not a whole number

9. Not a whole number

10. Whole number

11. Not a whole number

12. Not a whole number

Objective 2

13. 9, 4

14. 4, 6

15. 2, 0

16. 8, 1

17. 5, 3

18. 2, 0

19. 29, 176

20. 75, 229, 301

21. 70, 0, 603, 214

22. 300, 459, 200, 5

Objective 3

23. Eight thousand, seven hundred fourteen

24. Thirty-nine thousand, fifteen

25. Eight hundred thirty-four thousand, seven hundred sixty-eight

26. Two million, fifteen thousand, one hundred two

27. Ninety-six million, five hundred forty-three thousand, two hundred twenty-eight

28. Four hundred ninety-nine million, three hundred twenty-four thousand, five hundred eighteen

29. 4127

30. 29,516

31. 685,000,259

32. 300,075,002

33. 7210

34. 972,430

35. 15,313

36. 6,205,000

Objective 4

37. 177 calories   38. Running   39. 315 calories

40. 143 lbs

1.1 Mixed Exercises

41. Whole number   42. Not a whole number

43. Not a whole number   44. Whole number

45. Six thousand, two hundred forty-three   46. Nine hundred four

47. Sixteen thousand, two hundred one   48. 2321

49. 900,456   50. 7,006,012

51. Thousands   52. Tens

53. Hundred-thousands

1.2 Adding Whole Numbers

Objective 1

1. 8	2. 10	3. 12	4. 12	5. 13
6. 11	7. 16	8. 13	9. 11	10. 10
11. 14	12. 11	13. 10	14. 15	15. 15
16. 16	17. 9	18. 11	19. 14	20. 17

Objective 2

21. 25	22. 21	23. 19	24. 25	25. 25
26. 29	27. 17	28. 26	29. 33	30. 37
31. 31	32. 31	33. 26	34. 36	35. 28
36. 34	37. 29	38. 35	39. 34	40. 29

Objective 3

41. 69	42. 99	43. 78	44. 88	45. 97
46. 478	47. 988	48. 978	49. 895	50. 876
51. 52,879	52. 98,677	53. 69,776	54. 98,977	55. 58,596
56. 9989	57. 9899	58. 7798	59. 8698	60. 67,899

Objective 4

61. 81	62. 112	63. 218	64. 634	65. 1331
66. 1041	67. 603	68. 10,021	69. 49,580	70. 111,544
71. 60,011	72. 17,167	73. 15,815	74. 19,151	75. 9846
76. 10,415	77. 8096	78. 10,180	79. 3427	80. 5872

Objective 5

81. 38	82. 45	83. 44	84. 44
85. $50	86. 208 coins	87. 690 people	88. 625 tickets

89. 5231 books   90. 13,214 ft   91. 218 in   92. 310 ft

93. 763 meters   94. 1044 yd

Objective 6

95. Correct

96. Correct

97. Incorrect; 677

98. Incorrect; 1105

99. Correct

100. Incorrect; 18,430

101. Incorrect; 17,280

102. Correct

103. Incorrect; 1948

104. Correct

105. Correct

106. Incorrect; 9776

107. Correct

108. Incorrect; 11,822

109. Correct

110. Incorrect; 5585

1.2 Mixed Exercises

111. 13

112. 11

113. 15

114. 26

115. 18

116. 26

117. 99

118. 996

119. 125

120. 18,841

121. $1331

122. 73 coins

123. Correct

124. Correct

125. Incorrect; 9490

126. Correct

1.3 Subtracting Whole Numbers

Objective 1

1. $7 - 4 = 3, 7 - 3 = 4$

2. $15 - 6 = 9, 15 - 9 = 6$

3. $13 - 8 = 5, 13 - 5 = 8$

4. $23 - 15 = 8, 23 - 8 = 15$

5. $26 - 17 = 9, 26 - 9 = 17$

6. $32 - 19 = 13, 32 - 13 = 19$

7. $62 - 37 = 25, 62 - 25 = 37$

8. $112 - 89 = 23, 112 - 23 = 89$

9. $130 - 47 = 83, 130 - 83 = 47$

10. $82 - 47 = 35, 82 - 35 = 47$

11. $187 - 149 = 38, 187 - 38 = 149$

12. $312 - 253 = 59, 312 - 59 = 253$

13. $717 - 478 = 239, 717 - 239 = 478$

14. $1014 - 476 = 538, 1014 - 538 = 476$

15. $785 + 426 = 1211$

16. $183 + 604 = 787$

17. $117 + 87 = 204$

18. $47 + 266 = 313$

19. $2196 + 3721 = 5917$

20. $4981 + 113 = 5094$

Objective 2

21. 5, 3, 2

22. 7, 2, 5

23. 22, 7, 15

24. 35, 9, 24

24. 35, 9, 24

25. 36, 27, 9

26. 98, 36, 62

27. 18, 12, 6

28. 47, 29, 18

29. 187, 36, 151

30. 236, 142, 94

Objective 3

31. 31

32. 61

33. 22

34. 20

35. 21

36. 25

37. 151

38. 321

39. 310

40. 5151

41. 210

42. 4211

43. 2222

44. 2511

45. 6012

46. 12,222

Objective 4

47. Correct

48. Incorrect; 34

49. Correct

50. Incorrect; 8

51. Incorrect; 153

52. Incorrect; 249

53. Correct	54. Correct	55. Incorrect; 1079
56. Incorrect; 2980	57. Correct	58. Incorrect; 2560
59. Correct	60. Correct	61. Correct
62. Incorrect; 62,800	63. Correct	64. Correct
65. Incorrect; 78,087	66. Correct	

Objective 5

67. 7	68. 45	69. 29	70. 39
71. 25	72. 46	73. 192	74. 198
75. 178	76. 67	77. 1918	78. 25,899
79. 56,887	80. 43	81. 16	82. 243

Objective 6

83. 32 dogs	84. 25 boxes	85. 95 passengers	86. $419
87. 345 miles	88. 1796 people	89. 1134 people	90. $180
91. $5184	92. $76	93. $263	94. 63 miles
95. 2758 people			

1.3 Mixed Exercises

96. $57 - 26 = 31,$ $57 - 31 = 26$	97. $670 - 627 = 43$ $670 - 43 = 627$	98. $27 + 22 = 49$
99. $128 + 37 = 165$	100. 62, 17, 45	101. 424, 79, 345
102. 14, 121	103. 26, 231	104. 51, 111
105. 83,202	106. 39	107. 208
108. 14,662	109. 15,648	110. $105
111. $284	112. 76 athletes	

1.4 Multiplying Whole Numbers

Objective 1

1. Factors: 5, 3; product: 15
2. Factors: 4, 7; product: 28
3. Factors: 8, 4; product: 32
4. Factors: 5, 2; product: 10
5. Factors: 9, 8; product: 72
6. Factors: 14, 1; product: 14
7. Factors: 9,0; product: 0
8. Factors: 13, 3; product: 39
9. Factors: 7, 8; product: 56
10. Factors: 9, 12; product: 108
11. Factors: 17, 5; product: 85
12. Factors: 1, 5; product: 5

Objective 2

13. 32
14. 84
15. 8
16. 0
17. 0
18. 35
19. 192
20. 50
21. 120
22. 60

Objective 3

23. 40
24. 0
25. 48
26. 49
27. 21
28. 56
29. 54
30. 56
31. 36
32. 12
33. 63
34. 22
35. 60
36. 18
37. 216
38. 252
39. 228
40. 815
41. 2376
42. 6034
43. 4518
44. 1990
45. 4707
46. 2835
47. 2856
48. 14,028
49. 9024
50. 84,040
51. 180,054
52. 14,718
53. 31,350
54. 558,630
55. 115,413
56. 285,867

Objective 4

57. 820
58. 9100
59. 4290
60. 439,000
61. 25,560
62. 41,200
63. 8400
64. 523,800
65. 104,500
66. 6,010,000
67. 2,863,000
68. 485,840

69. 16,468,000   70. 4,350,500   71. 7,740,000

72. 20,000   73. 410,000   74. 282,000,000

75. 21,000   76. 4,800,000

Objective 5

77. 966   78. 1504   79. 864   80. 2964

81. 7448   82. 2349   83. 12,236   84. 11,684

85. 26,638   86. 11,043   87. 193,488   88. 492,119

89. 330,687   90. 379,745   91. 273,312   92. 1,129,351

Objective 6

93. 560 yd   94. 990 mi   95. $912   96. 672 cans

97. $756   98. $384   99. $1363   100. $36,105

101. $2800   102. $9728

1.4 Mixed Exercises

103. Factors: 3, 17; product: 51   104. Factors: 13, 3; product: 39

105. Factors: 0, 18; product: 0   106. Factors: 30, 1; product: 30

107. 60   108. 0   109. 140   110. 34

111. 1448   112. 40,800   113. 5700   114. 7120

115. 6,248,000   116. 1,934,996   117. 1,394,757   118. 31,869,992

119. 27,415,178 120. $14,178   121. $1602   122. $15,642

123. $3841

1.5 Dividing Whole Numbers

Objective 1

1. $\dfrac{15}{3} = 5,\ 3\overline{)15}^{\,5}$

2. $18 \div 6 = 3,\ 6\overline{)18}^{\,3}$

3. $27 \div 3 = 9,\ \dfrac{27}{3} = 9$

4. $39 \div 13 = 3,\ 13\overline{)39}^{\,3}$

5. $64 \div 8 = 8,\ \dfrac{64}{8} = 8$

6. $50 \div 25 = 2,\ 25\overline{)50}^{\,2}$

7. $28 \div 7 = 4,\ 7\overline{)28}^{\,4}$

8. $\dfrac{32}{16} = 2,\ 32 \div 16 = 2$

9. $42 \div 7 = 6,\ 7\overline{)42}^{\,6}$

10. $\dfrac{36}{12} = 3,\ 12\overline{)36}^{\,3}$

11. $0 \div 8 = 0,\ 8\overline{)0}^{\,0}$

12. $0 \div 24 = 0,\ \dfrac{0}{24} = 0$

13. $\dfrac{72}{12} = 6,\ 72 \div 12 = 6$

14. $25\overline{)100}^{\,4},\ 100 \div 25 = 4$

Objective 2

15. 27, 9, 3

16. 63, 7, 9

17. 30, 5, 6

18. 42, 14, 3

19. 38, 19, 2

20. 65, 13, 5

21. 28, 4, 7

22. 18, 9, 2

23. 52, 4, 13

24. 35, 7, 5

25. 132, 11, 12

26. 44, 11, 4

27. 63, 9, 7

28. 39, 13, 3

Objective 3

29. 0

30. 0

31. 0

32. 0

33. 0

34. 0

35. 0

36. 0

Objective 4

37. Undefined

38. Undefined

39. Undefined

40. Undefined

41. Undefined

42. Undefined

Objective 4

43. 1

44. 1

45. 1

46. 1

47. 1   48. 1   49. 1   50. 1

51. 1   52. 1

Objective 6

53. 17   54. 128   55. 38

56. 9   57. 27   58. 12

Objective 7

59. 12   60. 42   61. 13   62. 37

63. 62   64. 145   65. 115   66. 146

67. 170 R2   68. 144 R4   69. 41 R1   70. 141 R7

71. 120 R3   72. 204 R3   73. 72 R3   74. 164

Objective 8

75. Incorrect; 27   76. Correct   77. Incorrect; 1522 R5

78. Incorrect; 364 R3   79. Correct   80. Correct

81. Correct   82. Correct   83. Correct

84. Incorrect; 7133 R1   85. Correct   86. Incorrect; 5814

87. Correct   88. Incorrect; 15,763   89. Correct

90. Incorrect; 5012 R1   91. Incorrect; 6050   92. Correct

93. Incorrect; 355 R5   94. Incorrect; 3389 R7

Objective 9

95. Divisible by 2, 5, 10; not 3   96. Divisible by 2, 3; not 5, 10

97. Divisible by 3, 5; not 2, 10   98. Divisible by 2, 3; not 5, 10

99. Divisible by 3, 5; not 2, 10   100. Divisible by 3; not 2, 5, 10

101. Divisible by 3; not 2, 5, 10   102. Divisible by 2; not 3, 5, 10

103. Divisible by 2, 3, 5, 10   104. Divisible by 2, 3; not 5, 10

105. Divisible by 5; not 2, 3, 10

106. Divisible by 2, 5, 10; not 3

107. Divisible by 2, 5, 10; not 3

108. Divisible by 3; not 2, 5, 10

109. Divisible by 2, 3; not 5, 10

110. Not divisible by 2, 3, 5, 10

111. Divisible by 3, 5; not 2, 10

112. Divisible by 2, 3; not 5, 10

113. Divisible by 2, 5, 10; not 3

114. Divisible by 2; not 3, 5, 10

.5 Mixed Exercises

115. $\dfrac{42}{6} = 7,\ 6\overline{)42}^{\,7}$

116. $\dfrac{45}{15} = 3,\ 45 \div 15 = 3$

117. $8\overline{)128}^{\,16},\ 128 \div 8 = 16$

118. 45, 9, 5

119. 72, 9, 8

120. 27, 9, 3

121. Undefined

122. 0

123. Undefined

124. 1

125. 1

126. 1

127. 61 R3

128. 371 R4

129. 667 R5

130. 66 R6

131. True

132. True

133. True

134. False

1.6 Long Division

Objective 1

1. 82
2. 324
3. 77
4. 69

5. 102
6. 113 R34
7. 85 R84
8. 309 R1

9. 5814
10. 98,532 R6
11. 32,587
12. 300,959 R119

13. 654 R22
14. 5831

Objective 2

15. 9
16. 7
17. 9
18. 7

19. 6
20. 12
21. 16
22. 15

23. 70
24. 42
25. 80
26. 34

27. 84
28. 800

Objective 3

29. Correct
30. Correct
31. Correct

32. Incorrect; 346 R17
33. Incorrect; 45 R23
34. Correct

35. Incorrect; 296 R79
36. Incorrect; 460 R3
37. Correct

38. Correct

1.6 Mixed Exercises

39. 53
40. 207
41. 166 R18
42. 6354 R7

43. 6532
44. 4012
45. 80
46. 13

47. 27
48. 230
49. 9
50. 61

1.7 Rounding Whole Numbers

Objective 1

1. 8$\underline{5}$3

2. 1$\underline{0}$37

3. $\underline{4}$712

4. 6$\underline{4}$5,371

5. 4,$\underline{3}$16,214

6. 39,$\underline{9}$43,712

7. 643,$\underline{5}$19

8. $\underline{8}$1,243

9. 257,3$\underline{0}$1

10. 2,781,421

Objective 2

11. 7900

12. 450

13. 1380

14. 4940

15. 810

16. 18,200

17. 32,580

18. 9300

19. 53,600

20. 14,700

21. 8400

22. 41,100

23. 16,700

24. 5000

25. 4000

26. 52,000

Objective 3

27. $40 + 20 + 60 + 90 = 210$; 210

28. $20 + 90 + 40 + 20 = 170$; 161

29. $70 - 40 = 30$; 27

30. $90 - 50 = 40$; 36

31. $300 + 300 + 200 + 900 = 1700$; 1698

32. $400 + 200 + 300 + 200 = 1100$; 1125

33. $1000 - 400 = 600$; 589

34. $800 - 700 = 100$; 137

35. $900 \times 800 = 720,000$; 715,008

36. $900 \times 100 = 90,000$; 123,516

Objective 4

37. $600 + 40 + 200 + 2000 = 2840$; 3280

38. $700 + 30 + 700 + 80 = 1510$; 1549

39. $900 - 40 = 860$; 833

40. $300 - 50 = 250$; 264

41. $1000 \times 40 = 40,000$; 36,260

42. $400 \times 30 = 12,000$; 12,673

1.7 Mixed Exercises

43. 4$\underline{7}$99

44. 7$\underline{8}$2,563

45. 2$\underline{8}$,963,521

46. 54,000

47. 480,000

48. 600,000

49. 15,000,000

50. $800 \times 600 = 480,000$;
    $444,424$

51. $300 \times 400 = 120,000$;
    $119,498$

52. $700 \times 800 = 560,000$;
    $518,941$

53. $500 \times 100 = 50,000$;
    $64,218$

54. $3000 - 900 = 2100$;
    2354

55. $5000 - 300 = 4700$;
    4362

56. $700 \times 70 = 49,000$;
    $47,850$

57. $500 \times 80 = 40,000$;
    $42,201$

1.8 Exponents, Roots, and Order of Operations

Objective 1

1. 2, 7; 49
2. 2, 4; 16
3. 2, 9; 81
4. 3, 3; 27

5. 5, 1; 1
6. 4, 10; 10,000
7. 7, 2; 128
8. 3, 8; 512

Objective 2

9. 2
10. 3
11. 4
12. 11

13. 8
14. 10
15. 7
16. 13

17. 324; 324
18. 196; 14
19. 2500; 2500
20. 289; 17

21. 529; 529
22. 625; 625
23. 225; 225
24. 1296; 36

25. 400; 20
26. 2704; 2704

Objective 3

27. 39
28. 23
29. Undefined
30. 6

31. 27
32. 2
33. 54
34. Undefined

35. 45
36. 32
37. 75
38. 24

39. 28
40. 21

1.8 Mixed Exercises

41. 3, 5; 125
42. 3, 6; 216
43. 5, 3; 243
44. 20

45. 1
46. 7
47. 45
48. 8

49. 46
50. 89
51. 15
52. 6

1.9 Reading Pictographs, Bar Graphs, and Line Graphs

Objective 1

1. Georgia 2. Idaho 3. Minnesota 4. Georgia

Objective 2

5. 50 pints 6. 25 pints 7. Department 4 8. 25 pints

Objective 3

9. The nets dales are increasing every year.

10. $4,000,000,000

11. 2000

12. 2000

1.9 Mixed Exercises

13. French   14. 3   15. 12

16. Spanish   17. 250   18. 2900

19. 200   20. Freshmen   21. 1998, 1999, and 2000

22. 1998   23. 1996   24. $1.5 million

1.10 Reading Pictographs, Bar Graphs, and Lie Graphs

Objective 1

1. +

2. −

3. ×

4. +

5. ×

6. −

7. ÷

8. −

9. +

10. +

11. +

12. ÷

Objective 2

13. $1822

14. 4 toys

15. 4215 tickets

16. $642

17. $3008

18. $2900

19. 49 m

20. 828 salmon

21. 936 mi

22. $352

23. $268

24. 397 lb

Objective 3

25. $600 \div 60 = 10$ hr; 11 hr

26. $100 \div 50 = 2$ hr; 2 hr

27. $2000 \div 20 = \$100$; $79

28. $50,000 + 500 = \$50,500$; $47,541

29. $600 \div 20 = 30$ miles per gallon; 36 miles per gallon

30. $6000 \div 50 = \$120$; $115

31. $2000 - 1000 = \$1000$; $949

32. $4000 \div 40 = 100$ hr; 107 hr

33. $200,000 \div 30 = 6667$ people; 6850 people

34. $10,000 - (40 \times 200) = \$2000$; $3826

35. $10 + 200 + 400 = 610$ deer; 589 deer

36. $30,000 \times 30 = 900,000$ times; 983,040 times

37. $(2 \times 20) + (2 \times 3) = 46$ cookies; 36 cookies

38. $3000 - 300 - 700 - 200 = \$1800$; $1854

39. $10 \times 7 = 70$ forms; 84 forms

1.10 Mixed Exercises

40. −

41. ÷

42. ÷

43. ×

44. +

45. ×

46. ÷

47. ÷

48. 8230 lb

49. 4 strips

50. 217,800 sq ft

51. $275

52. 80,128,450 gal

53. 1011 vehicles

54. $6710

55. $552

56. $600 \div 20 = 30$ stamps; 35 stamps

57. $30,000 \div 300 = 100$ pages; 89 pages

58. $4000 - 900 = 3100$; $2825

59. $(200 \div 40) \times 5 = 25$ lb; 26 lb

60. $300 - 40 - 20 - 80 + 20 + 40 + 100 =$ 320 machines; 348 machines

Chapter 2

MULTIPLYING AND DIVIDING FRACTIONS

2.1 Basics of Fractions

Objective 1

1. $\frac{3}{8}$ 2. $\frac{1}{4}$ 3. $\frac{5}{6}$ 4. $\frac{2}{5}$ 5. $\frac{1}{3}$ 6. $\frac{5}{4}$

7. $\frac{5}{3}$ 8. $\frac{8}{5}$ 9. $\frac{5}{8}$ 10. $\frac{3}{2}$ 11. $\frac{1}{6}$ 12. $\frac{7}{12}$

Objective 2

13. N: 4; D: 3   14. N: 1; D: 2   15. N: 2; D: 5   16. N: 9; D; 23

17. N: 8; D: 11   18. N: 11; D: 8   19. N: 112; D: 5   20. N: 19; D: 50

21. N: 7; D: 15   22. N: 19; D: 8   23. N: 98; D: 13   24. N: 157; D: 12

Objective 3

25. Improper   26. Proper   27. Proper   28. Improper

29. Proper   30. Proper   31. Improper   32. Improper

33. Improper   34. Improper   35. Proper   36. Improper

37. Improper   38. Improper   39. Proper   40. Improper

2.1 Mixed Exercises

41. $\frac{2}{3}$ 42. $\frac{7}{10}$ 43. $\frac{3}{8}$ 44. $\frac{12}{7}$

45. N: 14; D: 195   46. N: 83; D: 85   47. N: 42; D: 23   48. N: 0; D: 16

49. Improper   50. Proper   51. Improper   52. Proper

2.2 Mixed Numbers

Objective 1

1. $2\frac{1}{2}, 1\frac{1}{6}$ 2. $5\frac{2}{3}, 3\frac{1}{2}$ 3. None

Objective 2

4. $\frac{23}{8}$ 5. $\frac{11}{6}$ 6. $\frac{14}{5}$ 7. $\frac{39}{7}$

8. $\frac{7}{4}$ 9. $\frac{25}{4}$ 10. $\frac{14}{3}$ 11. $\frac{15}{2}$

12. $\frac{29}{11}$ 13. $\frac{38}{7}$ 14. $\frac{20}{3}$ 15. $\frac{79}{9}$

Objective 3

16. $5\frac{1}{2}$ 17. $1\frac{3}{5}$ 18. $1\frac{1}{8}$ 19. $3\frac{3}{10}$

20. $1\frac{5}{9}$ 21. $2\frac{6}{7}$ 22. $3\frac{2}{9}$ 23. $3\frac{5}{7}$

24. $4\frac{1}{5}$ 25. $4\frac{5}{9}$ 26. $2\frac{7}{9}$ 27. $7\frac{1}{4}$

2.2 Mixed Exercises

28. $3\frac{1}{2}, 10\frac{1}{3}$ 29. $4\frac{3}{4}$ 30. $\frac{34}{3}$ 31. $\frac{37}{8}$

32. $\frac{43}{8}$ 33. $\frac{43}{5}$ 34. $\frac{31}{7}$ 35. $\frac{64}{9}$

36. $\frac{120}{9}$ 37. $\frac{250}{11}$ 38. $11\frac{3}{5}$ 39. $3\frac{6}{7}$

40. $4\frac{7}{10}$ 41. $4\frac{4}{13}$ 42. $30\frac{2}{3}$ 43. $19\frac{2}{11}$

44. $44\frac{1}{17}$ 45. $233\frac{10}{11}$

2.3 Factors

Objective 1

1. 1, 7

2. 1, 2, 3, 4, 6, 12

3. 1, 7, 49

4. 1, 3, 5, 15

5. 1, 2, 5, 10

6. 1, 2, 3, 4, 6, 9, 12, 18, 36

7. 1, 5, 25

8. 1, 2, 3, 4, 6, 8, 12, 24

9. 1, 2, 3, 6, 9, 18

10. 1, 2, 3, 5, 6, 10, 15, 30

11. 1, 2, 3, 4, 6, 8, 9, 12, 18, 24, 36, 72

12. 1, 2, 4, 7, 14, 28

Objective 2

13. Neither

14. Prime

15. Prime

16. Composite

17. Composite

18. Composite

19. Prime

20. Prime

21. Prime

22. Composite

23. Composite

24. Composite

25. Composite

26. Prime

27. Composite

28. Composite

Objective 3

29. $2^2 \cdot 3$

30. $2 \cdot 11$

31. $3 \cdot 5$

32. 3^3

33. $2^2 \cdot 7$

34. $2 \cdot 3 \cdot 7$

35. 2^5

36. $2^3 \cdot 3$

37. $3^2 \cdot 7$

38. $2^2 \cdot 5^2$

39. $2^3 \cdot 3^2$

40. $2^3 \cdot 7$

2.3 Mixed Exercises

41. 1, 2, 4, 5, 8, 10, 20, 40

42. 1, 2, 3, 6, 11, 22, 33, 66

43. Composite

44. Composite

45. Prime

46. Composite

47. $2 \cdot 5 \cdot 7$

48. $2^2 \cdot 3^3$

49. $5 \cdot 17$

50. $2^5 \cdot 5^2$

51. $2^4 \cdot 3 \cdot 5$

52. $2^5 \cdot 5$

53. $2 \cdot 3^2 \cdot 5^2$

54. $3^2 \cdot 19$

2.4 Writing a Fraction in Lowest Terms

Objective 1

1. No   2. Yes   3. No   4. Yes   5. No   6. No

7. Yes   8. No   9. No   10. Yes   11. No   12. No

Objective 2

13. $\frac{1}{4}$ 14. $\frac{1}{3}$ 15. $\frac{2}{7}$ 16. $\frac{2}{9}$

17. $\frac{5}{6}$ 18. $\frac{5}{9}$ 19. $\frac{2}{7}$ 20. $\frac{2}{7}$

Objective 3

21. $\frac{3}{4}$ 22. $\frac{1}{2}$ 23. $\frac{6}{7}$ 24. $\frac{4}{5}$

25. $\frac{2}{3}$ 26. $\frac{1}{2}$ 27. $\frac{3}{20}$ 28. $\frac{8}{11}$

Objective 4

29. Not equivalent   30. Equivalent   31. Equivalent

32. Not equivalent   33. Not equivalent   34. Not equivalent

35. Equivalent   36. Equivalent   37. Equivalent

38. Not equivalent   39. Not equivalent   40. Equivalent

2.4 Mixed Exercises

41. Yes   42. No   43. Yes   44. Yes

45. $\frac{3}{22}$ 46. $\frac{5}{7}$ 47. $\frac{2}{7}$ 48. $\frac{2}{3}$

49. $\frac{2}{5}$ 50. $\frac{5}{6}$ 51. $\frac{5}{12}$ 52. $\frac{23}{49}$

53. Equivalent   54. Not equivalent   55. Equivalent   56. Not equivalent

57. Not equivalent   58. Equivalent   59. Not equivalent   60. Equivalent

2.5 Multiplying Fractions

Objective 1

1. $\frac{5}{24}$ 2. $\frac{35}{54}$ 3. $\frac{2}{15}$ 4. $\frac{12}{35}$ 5. $\frac{55}{24}$ 6. $\frac{27}{20}$

7. $\frac{7}{40}$ 8. $\frac{8}{7}$ 9. $\frac{2}{3}$ 10. $\frac{3}{8}$ 11. $\frac{5}{81}$ 12. $\frac{1}{96}$

Objective 2

13. $\frac{3}{16}$ 14. $\frac{1}{4}$ 15. $\frac{5}{12}$ 16. $\frac{5}{9}$ 17. $\frac{2}{3}$ 18. $\frac{3}{5}$

19. $\frac{2}{21}$ 20. $\frac{30}{7}$ 21. $\frac{1}{4}$ 22. $\frac{5}{9}$ 23. $\frac{1}{6}$ 24. $\frac{5}{12}$

Objective 3

25. 15   26. 42   27. 45   28. 4   29. 18   30. $3\frac{1}{2}$

31. 7   32. $\frac{7}{8}$ 33. 5

Objective 4

34. $\frac{1}{6}$ square foot   35. $\frac{1}{3}$ square yard   36. $\frac{3}{8}$ square meter

37. $\frac{3}{22}$ square inch   38. $2\frac{1}{2}$ square yards   39. $\frac{16}{99}$ square foot

40. $\frac{21}{256}$ square inch   41. $\frac{15}{32}$ square meter   42. $\frac{5}{9}$ square yard

43. $\frac{25}{36}$ square yard

2.5 Mixed Exercises

44. $\frac{5}{12}$ 45. $\frac{7}{15}$ 46. $\frac{1}{2}$ 47. $\frac{9}{70}$

48. $\frac{8}{9}$ 49. $\frac{3}{4}$ 50. $\frac{1}{6}$ 51. $\frac{1}{7}$

52. 3   53. $\frac{7}{50}$ 54. $\frac{4}{5}$ 55. $1\frac{1}{2}$

56. $1\frac{5}{7}$ 57. $4\frac{2}{3}$ 58. 9   59. $\frac{6}{91}$ square yard

60. $\frac{5}{8}$ square feet   61. $\frac{1}{4}$ square inch

2.6 Applications of Multiplication

Objective 1

1. 1680 paperbacks
2. 500 items
3. $1500

4. $30
5. 133 employees
6. 228 students

7. $500
8. 27 miles
9. $12,000

10. $14,400
11. $2250

2.7   Dividing Fractions

Objective 1

1. $\frac{4}{3}$

2. $\frac{2}{9}$

3. 3

4. $\frac{7}{6}$

5. $\frac{1}{10}$

6. $\frac{4}{15}$

Objective 2

7. $\frac{1}{3}$

8. $2\frac{2}{15}$

9. $\frac{1}{4}$

10. $\frac{3}{4}$

11. $\frac{16}{25}$

12. $3\frac{1}{3}$

13. 6

14. $37\frac{1}{2}$

15. $\frac{1}{24}$

16. $\frac{1}{33}$

17. $\frac{11}{15}$

18. $2\frac{1}{3}$

Objective 3

19. $\frac{7}{32}$ acres

20. 54 dresses

21. 40 salt shakers

22. 24 Brownies

23. 63 vials

24. 32 guests

25. 90 bows

2.7   Mixed Exercises

26. $\frac{5}{8}$

27. $\frac{11}{4}$

28. $\frac{1}{12}$

29. $7\frac{9}{13}$

30. 75

31. $12\frac{4}{7}$

32. $20\frac{2}{5}$

33. 24 patties

34. 40 trips

35. 16 tumblers

2.8 Multiplying and Dividing Mixed Numbers

Objective 1

1. $5 \cdot 3 = 15; 13\frac{1}{3}$
2. $4 \cdot 4 = 16; 15$
3. $5 \cdot 3 = 15; 16\frac{4}{5}$

4. $4 \cdot 1 = 4; 5$
5. $4 \cdot 2 = 8; 10\frac{2}{3}$
6. $6 \cdot 7 = 42; 40\frac{3}{8}$

7. $2 \cdot 4 = 8; 8\frac{1}{8}$
8. $18 \cdot 3 = 54; 46$
9. $6 \cdot 4 = 24; 21$

10. $3 \cdot 15 = 45; 51$
11. $1 \cdot 3 \cdot 2 = 6; 5$
12. $1 \cdot 1 \cdot 2 = 2; 2\frac{1}{2}$

Objective 2

13. $6 \div 5 = 1\frac{1}{5}; 1\frac{1}{9}$
14. $3 \div 3 = 1; 1\frac{3}{22}$
15. $5 \div 1 = 5; 3\frac{7}{10}$

16. $4 \div 4 = 1; 1\frac{1}{4}$
17. $6 \div 1 = 6; 4\frac{4}{5}$
18. $6 \div 3 = 2; 2\frac{1}{2}$

19. $5 \div 4 = 1\frac{1}{4}; 1\frac{1}{3}$
20. $14 \div 8 = 1\frac{3}{4}; 1\frac{2}{3}$
21. $7 \div 6 = 1\frac{1}{6}; 1\frac{2}{9}$

22. $5 \div 2 = 2\frac{1}{2}; 2\frac{1}{3}$
23. $8 \div 1 = 8; 11\frac{1}{4}$
24. $3 \div 2 = 1\frac{1}{2}; 1\frac{1}{2}$

Objective 3

25. $20 \cdot 3 = 60$ yards; 65 yards
26. $38 \cdot 9 = \$342; \$344.25$

27. $41 \cdot 6 = 246$ yards; 248 yards
28. $66 \div 6 = 11$ acres; $11\frac{2}{7}$ acres

29. $29 \cdot 2 = 58$ ounces; $50\frac{2}{5}$ ounces
30. $15 \div 1 = 15$ cans; 20 cans

2.8 Mixed Exercises

31. $9 \cdot 3 \cdot 1 \cdot 3 = 81; 96$
32. $3 \cdot 3 \cdot 1 = 9; 14\frac{7}{8}$

33. $4 \cdot 2 \cdot 3 = 24; 26\frac{2}{3}$
34. $3 \cdot 5 \cdot 6 = 90; 85$

35. $3 \div 2 = 1\frac{1}{2}; 1\frac{9}{10}$
36. $9 \div 5 = 1\frac{4}{5}; 1\frac{3}{4}$

37. $16 \div 3 = 5\frac{1}{3}; 5\frac{13}{23}$
38. $6 \div 4 = 1\frac{1}{2}; 1\frac{1}{2}$

39. $70 \div 4 = 17\frac{1}{2}$ dresses; 16 dresses
40. $25 \cdot 9 = \$225; \$220.50$

41. $4 \cdot 0 = 0$ pounds; $\frac{7}{8}$ pound
42. $8 \div 1 = 8$ tapes; 6 tapes

Chapter 3

ADDING AND SUBTRACTING FRACTIONS

3.1 Adding and Subtracting Like Fractions

Objective 1

1. Like   2. Like   3. Unlike   4. Unlike   5. Like

6. Unlike   7. Unlike   8. Unlike   9. Unlike   10. Like

11. Like   12. Like

Objective 2

13. 1   14. $\frac{5}{7}$ 15. $\frac{3}{4}$ 16. $\frac{4}{5}$

17. $1\frac{1}{2}$ 18. $3\frac{2}{3}$ 19. $1\frac{1}{8}$ 20. $1\frac{1}{4}$

21. 1   22. $1\frac{2}{5}$ 23. $1\frac{3}{10}$ 24. $\frac{4}{7}$

25. $\frac{7}{11}$ 26. $\frac{1}{2}$ mile

Objective 3

27. $\frac{8}{13}$ 28. $\frac{1}{5}$ 29. 1   30. $\frac{1}{2}$

31. $\frac{1}{2}$ 32. $\frac{2}{3}$ 33. $\frac{2}{3}$ 34. $\frac{3}{25}$

35. $\frac{4}{5}$ 36. $\frac{5}{14}$ 37. $\frac{5}{9}$ 38. $\frac{1}{6}$

39. $\frac{2}{3}$ mile 40. $\frac{2}{5}$

3.1 Mixed Exercises

41. Unlike   42. Like   43. Like   44. Unlike

45. 2   46. $\frac{2}{3}$ 47. $1\frac{1}{3}$ 48. $\frac{1}{2}$

49. $\frac{3}{5}$ 50. $\frac{3}{7}$ 51. $\frac{2}{3}$ 52. $\frac{1}{3}$

3.2 Least Common Multiples

Objective 1

1. 14	2. 16	3. 36	4. 42
5. 84	6. 140	7. 260	8. 200
9. 105			

Objective 2

10. 60	11. 80	12. 24	13. 75
14. 90	15. 70	16. 18	17. 84
18. 160			

Objective 3

19. 28	20. 144	21. 224	22. 175
23. 480	24. 180	25. 400	26. 420
27. 840			

Objective 4

28. 12	29. 18	30. 30	31. 110
32. 288	33. 130	34. 595	35. 108
36. 72	37. 120	38. 140	39. 180

Objective 5

40. 4	41. 14	42. 4	43. 18
44. 20	45. 48	46. 36	47. 90
48. 6	49. 60	50. 33	51. 12
52. 35	53. 28	54. 105	

3.2 Mixed Exercises

55. 75	56. 56	57. 90	58. 60

59. 60   60. 72   61. 144   62. 180

63. 3900   64. 756   65. 1680   66. 675

67. 6300

In Exercises 68 – 72 only the new numerators are given.

68. 180   69. 126   70. 95   71. 108

72. 54

3.3   Adding and Subtracting Unlike Fractions

Objective 1

1. $\frac{5}{6}$

2. $\frac{5}{6}$

3. $\frac{33}{40}$

4. $\frac{21}{26}$

5. $\frac{23}{30}$

6. $\frac{7}{8}$

7. $\frac{1}{3}$

8. $\frac{47}{48}$

9. $\frac{37}{45}$

10. $\frac{19}{24}$

11. $\frac{19}{24}$

12. $\frac{17}{18}$

Objective 2

13. $\frac{5}{6}$

14. $\frac{23}{24}$

15. $\frac{5}{6}$

16. $\frac{7}{18}$

17. $\frac{13}{21}$

18. $\frac{29}{66}$

19. $\frac{3}{4}$

20. $\frac{19}{42}$

Objective 3

21. $\frac{3}{8}$

22. $\frac{19}{36}$

23. $\frac{1}{2}$

24. $\frac{11}{24}$

25. $\frac{1}{3}$

26. $\frac{1}{3}$

27. $\frac{5}{24}$

28. $\frac{5}{36}$

29. $\frac{1}{24}$

30. $\frac{1}{15}$

31. $\frac{1}{6}$

32. $\frac{1}{6}$

33. $\frac{7}{24}$

34. $\frac{11}{24}$

3.3   Mixed Exercises

35. $\frac{8}{21}$

36. $\frac{43}{60}$

37. $\frac{3}{4}$

38. $\frac{7}{8}$

39. $\frac{29}{54}$

40. $\frac{5}{24}$

41. $\frac{34}{45}$

42. $\frac{1}{36}$

43. $\frac{21}{80}$

44. $\frac{5}{8}$

45. $\frac{7}{20}$

46. $\frac{37}{50}$

3.4 Adding and Subtracting Mixed Numbers

Objective 1

1. $5+4=9$; $9\frac{4}{7}$

2. $3+5=8$; $7\frac{71}{72}$

3. $8+5=13$; $12\frac{3}{8}$

4. $18+12+6=36$; $35\frac{17}{24}$

5. $60+25+13=98$; $97\frac{19}{24}$

6. $127+29+13=169$; $168\frac{5}{6}$

7. $6-2=4$; $3\frac{1}{2}$

8. $13-2=11$; $10\frac{3}{16}$

9. $10-2=8$; $7\frac{1}{4}$

10. $2+2=4$ cans; $4\frac{5}{24}$ cans

11. $9+2=11$ gallons; $11\frac{1}{12}$ gallons

12. $2+3=5$ boxes; $5\frac{1}{24}$ boxes

13. $8+10=18$ tons; $17\frac{7}{20}$ tons

14. $13-6=7$ hours; $6\frac{3}{8}$ hours

15. $6+8+9+8+8=39$ hours;
$38\frac{9}{20}$ hours

15. $6+8+9+8+8=39$ hours;
$38\frac{9}{20}$ hours

16. $5+9+3+1=18$ tons; $17\frac{1}{3}$ tons

Objective 2

17. $9-7=2$; $1\frac{3}{4}$

18. $11-7=4$; $4\frac{1}{2}$

19. $6-6=0$; $\frac{3}{4}$

20. $7-6=1$; $1\frac{17}{20}$

21. $27-19=8$; $8\frac{13}{30}$

22. $12-12=0$; $\frac{35}{48}$

23. $43-30=13$; $12\frac{23}{24}$

24. $42-20=22$; $22\frac{1}{4}$

25. $21-18=3$; $3\frac{7}{16}$

26. $40-8-6-8-9=9$ hours; $8\frac{7}{8}$ hours

27. $146-35-43-33=35$ yards;
$34\frac{5}{8}$ yards

28. $12-1-2-3=6$ cu yards; $5\frac{1}{24}$ cu yards

29. $15-3-5-4=3$ yards; $3\frac{1}{4}$ yards

30. $2\frac{3}{16}$

31. $15\frac{1}{20}$

Objective 3

32. $5\frac{1}{4}$ 33. $8\frac{1}{6}$ 34. $3\frac{39}{40}$ 35. $5\frac{7}{24}$

36. $2\frac{3}{8}$ 37. $1\frac{5}{6}$ 38. $2\frac{7}{8}$ 39. $1\frac{5}{6}$

40. $1\frac{17}{24}$

3.4 Mixed Exercises

41. $23+16=39;\ 38\frac{2}{5}$ 42. $15+9=24;\ 23\frac{2}{3}$

43. $10+9=19;\ 18\frac{3}{8}$ 44. $27+15+10=52;\ 52\frac{1}{8}$

45. $29+48+24=101;\ 99\frac{5}{6}$ 46. $29+21+20=70;\ 69\frac{17}{20}$

47. $8-7=1;\ 1\frac{1}{24}$ 48. $16-9=7;\ 7\frac{19}{48}$

49. $27-13=14;\ 13\frac{32}{63}$ 50. $373-208=165;\ 164\frac{11}{24}$

51. $9-7=2;\ 1\frac{49}{72}$ 52. $15-8=7;\ 6\frac{25}{56}$

53. $15-12=3;\ 2\frac{8}{15}$ 54. $29-8=21;\ 20\frac{11}{12}$

55. $130-99=31;\ 30\frac{11}{15}$ 56. $147-40=107;\ 107\frac{1}{6}$

57. $28-4=24;\ 23\frac{4}{7}$ 58. $75-1=74;\ 74\frac{3}{8}$

3.5 Order Relations and the Order of Operations

Objective 1

1. < 2. < 3. > 4. >

5. < 6. > 7. < 8. >

9. >

Objective 2

10. $\frac{1}{64}$ 11. $\frac{1}{4}$ 12. $\frac{4}{9}$ 13. $\frac{1}{25}$

14. $4\frac{17}{27}$ 15. $\frac{1}{125}$ 16. $\frac{1}{16}$ 17. $5\frac{1}{16}$

18. $\frac{27}{343}$ 19. $\frac{27}{64}$ 20. $\frac{64}{121}$ 21. $\frac{64}{225}$

22. $\frac{8}{729}$ 23. $2\frac{46}{49}$ 24. $4\frac{52}{243}$

Objective 3

25. $2\frac{2}{3}$ 26. $\frac{1}{16}$ 27. $\frac{4}{15}$ 28. $\frac{2}{15}$

29. $\frac{4}{25}$ 30. $\frac{1}{36}$ 31. $\frac{1}{28}$ 32. $\frac{9}{16}$

33. $1\frac{5}{12}$ 34. $\frac{5}{8}$ 35. $\frac{5}{28}$ 36. $2\frac{1}{4}$

37. $1\frac{3}{5}$ 38. $\frac{1}{2}$ 39. $\frac{1}{8}$

3.5 Mixed Exercises

40. < 41. > 42. < 43. $\frac{5}{9}$

44. $\frac{7}{81}$ 45. $\frac{1}{4}$ 46. $\frac{33}{40}$ 47. $1\frac{1}{6}$

48. 2 49. $\frac{15}{16}$ 50. $\frac{19}{144}$ 51. $\frac{3}{40}$

52. $\frac{1}{4}$

Chapter 4

WHOLE NUMBERS

4.1 Reading and Writing Decimals

Objective 1

1. $\frac{2}{10}$; 0.2; two tenths

2. $\frac{5}{10}$; 0.5; five tenths

3. $\frac{8}{10}$; 0.8; eight tenths

4. $\frac{19}{100}$; 0.19; nineteen hundredths

5. $\frac{35}{100}$; 0.35; thirty-five hundredths

6. $\frac{3}{100}$; 0.03; three hundredths

7. $\frac{58}{100}$; 0.58; fifty-eight hundredths

8. $\frac{47}{100}$; 0.47; forty-seven hundredths

Objective 2

9. 5, 0

10. 4, 9

11. 6, 3

12. 4, 7

13. 6, 9

14. 2, 5

15. 7, 1

16. 2, 5

17. 4, 6

18. 6, 4

19. 3, 6

20. Tenths, hundredths

21. Tenths, hundredths

22. Tenths, hundredths

23. Tenths, hundredths

24. Tenths, hundredths, thousandths

25. Tenths, hundredths thousandths

Objective 3

26. Eight hundredths

27. Seven thousandths

28. Four and six hundredths

29. Three and fourteen ten-thousandths

30. Five hundred sixty-one ten-thousandths

31. Ten and eight hundred thirty-five thousandths

32. Two and three hundred four thousandths

33. Ninety-seven and eight thousandths

34. 5.04

35. 11.009

36. 38.00052

37. 300.0023

Objective 4

38. $\frac{4}{5}$

39. $\frac{1}{10}$

40. $3\frac{3}{5}$

41. $\frac{1}{2}$

42. $4\frac{13}{50}$

43. $\frac{19}{20}$

44. $1\frac{33}{50}$

45. $\frac{99}{100}$

46. $3\frac{3}{4}$

4.1 Mixed Exercises

47. $\frac{43}{100}$; 0.43; forty-three hundredths

48. $\frac{8}{100}$; 0.08; eight hundredths

49. 5, 2

50. 2, 3

51. 2, 7

52. Tenths, hundredths, thousandths

53. Tenths, hundredths, thousandths

54. Tens, ones, tenths, hundredths

55. Tens, ones, tenths, hundredths, thousandths

56. Four and eighty-three thousandths

57. Forty-nine thousandths

58. 10.0006

59. 3612.017

60. $\frac{89}{100}$

61. $\frac{1}{25}$

62. $4\frac{2}{25}$

63. $\frac{63}{125}$

64. $6\frac{93}{200}$

4.2 Rounding Decimals

Objective 2

1. 17.9

2. 489.8

3. 785.498

4. 43.5150

5. 53.33

6. 75.4

7. 89.53; 89.5

8. 21.77; 21.8

9. 0.89; 0.9

10. 1.44; 1.4

11. 0.10; 0.1

12. 114.04; 114.0

13. 101.75; 101.7

14. 78.70; 78.7

15. 108.07; 108.1

Objective 3

16. $79

17. $28

18. $226

19. $4798

20. $11,840

21. $27,870

22. $1.25

23. $1.09

24. $112.01

25. $134.21

26. $1028.67

27. $2096.01

4.2 Mixed Exercises

28. 39.910

29. 10.32

30. 799.8

31. 486

32. 3257.60; 3257.6

33. 486.93; 486.9

34. 264.99; 265.0

35. 304.86; 304.9

36. 27.57; 27.6

37. $55

38. $276

39. $2461

40. $62.18

41. $495.62

42. $1.49

4.3 Adding and Subtracting Decimals

Objective 1

1. 92.49
2. 177.951
3. 105.43

4. 178
5. 72.453
6. 15.45

7. 48.35
8. 38.4133 centimeters
9. 123.6802 inches

Objective 2

10. 66.5
11. 115.8
12. 32.44

13. 42.566
14. 16.124
15. 58.32

16. 609.168
17. 24.016 feet
18. 2.758 yards

Objective 3

19. $30+40+8=78$; 82.91
20. $20-8=12$; 13.16

21. $10-5=5$; 4.838
22. $600+30+50=680$; 676.60

23. $9-3=6$; 5.53
24. $30+30+20=\$80$; \$75.57

25. $400+90=\$490$; 452.88
26. $30-10=20$ hours; 17.85 hours

27. $4+5+5=14$ days; 13.4 days
28. $20-10=\$10$; \$8.71

29. $50-40=\$10$; \$12.43
30. $9+10+30=\$49$; \$47.92

31. $80-50=\$30$; \$29.97
32. $80,000+80+80=80,160$ miles; 80,611.3 miles

4.3 Mixed Exercises

33. 10.63
34. 10.982
35. 64.706
36. 92.335

37. 1526.5
38. 26.969
39. 22.837
40. 33.357

41. 331.966 yards
42. 1186.1162 feet
43. 0.862 inch
44. 23.038 feet

45. $50+400+5=455$; 485.019
46. $50-20=30$; 33.62

47. $100+30+60=190$; 233
48. $300+500+300=1100$; 1050.76

49. $600 - 90 = 510$; 494.516

50. $900 - 30 = 870$; 862.482

51. $200 + 300 + 300 = 800$ miles;
     779.7 miles

52. $10 + 3 + 10 + 2 + 10 = 35$ hours;
     39.7 hours

53. $90,000 - 90,000 = 0$ miles;
     691.9 miles

54. $40,000 - 30,000 = 10,000$ miles;
     12,474.9 miles

55. $1000 + 200 + 400 + 8000 = \$9600$;
     \$9549.69

56. $1000 + 700 + 1000 - 1000 - 200 = \$1500$;
     \$1274.07

4.4 Multiplying Decimals

Objective 1

1. 0.2279

2. 2.6598

3. 90.71

4. 443.9

5. 1.5548

6. 73.386

7. 0.0037

8. 0.0000126

9. 0.00000963

10. $163.08

11. $310.63

12. $9.44

13. $7.35

Objective 2

14. $50 \times 6 = 300$; 288.26

15. $30 \times 3 = 90$; 101.32

16. $60 \times 4 = 240$; 218.4756

17. $30 \times 20 = 600$; 756.6478

18. $80 \times 1 = 80$; 43.548

19. $3 \times 4 = 12$; 10.465

20. $400 \times 8 = 3200$;
    3033.306

21. $30 \times 10 = 300$; 308

22. $494.40

23. $14,858.88

24. $2105.99

25. $25.03

4.4 Mixed Exercises

26. 2.01344

27. 2.15924

28. 78.02

29. 0.0000477

30. 0.00031

31. $200.20

32. $199.00

33. $20.64

34. $7.25

35. $100 \times 10 = 1000$;
    1417.3383

36. $300 \times 9 = 2700$;
    2742.762

37. $476.29

38. $2707.50

4.5 Dividing Decimals

Objective 1

1. 14.65   2. 1.794   3. 2.359   4. 6.968   5. 7.994

6. 2.838   7. 4.271   8. 6.156   9. 20.9   10. 28.079

11. 17.718   12. 16.589   13. $1.99 per   14. $0.80 per
                                sock   pound

Objective 2

15. 129.467   16. 3.796   17. 130.983

18. 116.9   19. 17.660   20. 128.25

21. 234.332   22. 9.549   23. 67.231

24. 53,950.943   25. 162.791   26. 33.3 miles per gallon

27. $3.55 per yard   28. $0.32 per brick 29. $0.16 per paper

30. $6.71 per hour

Objective 3

31. Reasonable   32. Reasonable   33. Unreasonable

34. Unreasonable   35. Reasonable   36. Reasonable

37. Unreasonable   38. Unreasonable   39. Unreasonable

40. Reasonable

Objective 4

41. 20.31   42. 13.51   43. 54.02   44. 16.549   45. 96.61

46. 53.24   47. 7.91   48. 5.6   49. 53.548   50. 13.49

4.5 Mixed Exercises

52. 1.189   52. 0.492   53. 0.845

54. 4.350   55. $3.16 per book   56. $0.05 per pencil

57. 0.866   58. 103   59. 2907

60. 163.334

61. 131.061

62. $0.97 per record

63. 21 months

64. 43 months

65. Unreasonable

66. Reasonable

67. Unreasonable

68. 5.54

69. 44.91

70. 51.7

4.6 Writing Fractions as Decimals

Objective 1

1. 6.5   2. 0.2   3. 2.667   4. 0.125

5. 0.091   6. 7.1   7. 0.6   8. 0.875

9. 4.111   10. 0.52   11. 0.15   12. 31.231

Objective 2

13. <   14. <   15. >

16. <   17. >   18. >

19. $\frac{3}{11}$, 0.29, $\frac{1}{3}$ 20. 0.88, $\frac{8}{9}$, 0.89   21. 0.166, 0.1666, $\frac{1}{6}$

22. 0.466, $\frac{7}{15}$, $\frac{9}{19}$

4.6 Mixed Exercises

23. 0.467   24. 0.417   25. 0.611

26. 19.708   27. >   28. <

29. >   30. <   31. $\frac{1}{7}$, 0.187, $\frac{3}{16}$

32. $\frac{11}{13}$, 0.8462, $\frac{6}{7}$

Chapter 5

RATIO AND PROPORTION

5.1 Ratios

Objective 1

1. $\frac{7}{8}$
2. $\frac{3}{4}$
3. $\frac{76}{101}$
4. $\frac{5}{14}$
5. $\frac{19}{25}$

6. $\frac{8}{3}$
7. $\frac{17}{27}$
8. $\frac{1}{4}$
9. $\frac{7}{3}$
10. $\frac{3}{64}$

Objective 2

11. $\frac{13}{4}$
12. $\frac{11}{8}$
13. $\frac{6}{5}$
14. $\frac{9}{2}$
15. $\frac{5}{6}$

16. $\frac{2}{1}$
17. $\frac{3}{4}$
18. $\frac{31}{44}$
19. $\frac{8}{5}$
20. $\frac{3}{2}$

Objective 3

21. $\frac{2}{7}$
22. $\frac{16}{5}$
23. $\frac{9}{5}$
24. $\frac{7}{2}$

25. $\frac{5}{6}$
26. $\frac{4}{15}$
27. $\frac{5}{8}$
28. $\frac{2}{1}$

5.1 Mixed Exercises

29. $\frac{25}{14}$
30. $\frac{13}{27}$
31. $\frac{35}{1}$
32. $\frac{5}{2}$
33. $\frac{21}{20}$

34. $\frac{2}{3}$
35. $\frac{1}{15}$
36. $\frac{2}{3}$
37. $\frac{1}{6}$
38. $\frac{8}{1}$

5.2 Rates

Objective 1

1. $\dfrac{3 \text{ miles}}{1 \text{ minute}}$

2. $\dfrac{5 \text{ feet}}{1 \text{ second}}$

3. $\dfrac{7 \text{ dresses}}{1 \text{ person}}$

4. $\dfrac{5 \text{ horses}}{1 \text{ team}}$

5. $\dfrac{15 \text{ gallons}}{1 \text{ hour}}$

6. $\dfrac{15 \text{ miles}}{1 \text{ gallon}}$

7. $\dfrac{7 \text{ pills}}{1 \text{ patient}}$

8. $\dfrac{9 \text{ kilometers}}{1 \text{ liter}}$

9. $\dfrac{32 \text{ pages}}{1 \text{ chapter}}$

10. $\dfrac{55 \text{ miles}}{1 \text{ hour}}$

Objective 2

11. $15/hour

12. $175/day

13. $110/day

14. $225/pound

15. $13.64/hour

16. 20.2 miles/gallon

17. $\frac{1}{2}$ crate/minute; 2 minutes/crate

18. $\frac{1}{2}$ acre/hour; 2 hour/acre

19. $9.18/hour

20. $5.90/yard

Objective 3

21. 16 ounces for $1.89

22. 16 ounces for $0.89

23. 12 cans for $3.59

24. 5 cans for $2.75

5.2 Mixed Exercises

25. 27.3 miles/gallons

26. $\frac{1}{4}$ pound/person

27. $8.75/hour

28. $103.30/day

29. $35.90/square yard

30. $11.50/share

31. $2.58/share

32. 20 oz for $2.29

5.3 Proportions

Objective 1

1. $\frac{11}{15} = \frac{22}{30}$ 2. $\frac{50}{8} = \frac{75}{12}$ 3. $\frac{24}{30} = \frac{8}{10}$ 4. $\frac{36}{45} = \frac{8}{10}$ 5. $\frac{14}{21} = \frac{10}{15}$

6. $\frac{3}{33} = \frac{12}{132}$ 7. $\frac{26}{4} = \frac{39}{6}$ 8. $\frac{9}{3} = \frac{42}{14}$ 9. $\frac{1\frac{1}{2}}{4} = \frac{21}{56}$ 10. $\frac{3\frac{2}{3}}{11} = \frac{10}{30}$

Objective 2

11. True   12. False   13. False   14. True   15. False

16. False   17. True   18. True   19. False   20. False

21. False   22. True

Objective 3

23. False   24. True   25. True   26. False   27. False

28. True   29. True   30. False   31. False   32. True

33. True   34. False

5.3 Mixed Exercises

35. $\frac{6}{21} = \frac{10}{35}$ 36. $\frac{9}{15} = \frac{21}{35}$ 37. False   38. True   39. True

40. True   41. True   42. False   43. False   44. False

5.4 Solving Proportions

Objective 1

1. 9	2. 16	3. 36	4. 4
5. 21	6. 20	7. 5	8. 70
9. 45	10. 48	11. 77	12. 12

Objective 2

13. $1\frac{3}{4}$	14. $11\frac{2}{3}$	15. $2\frac{2}{3}$	16. 13
17. 1	18. $4\frac{1}{2}$	19. $2\frac{1}{2}$	20. 0
21. 0	22. 21	23. $1\frac{1}{2}$	24. 14

5.4 Mixed Exercises

25. 18	26. 40	27. 55	28. 33	29. 3	30. $\frac{1}{15}$	31. 8
32. 0	33. 6					

5.5 Solving Application Problems with Proportions

Objective 1

1. $112.50

2. 12 inches

3. $15

4. $108

5. 8 pounds

6. $259.20

7. $818.40

8. 55 ounces

9. 960 miles

10. 12 inches

11. $440

12. $2.40

13. $4399.50

14. $768,000

15. $122.50

16. $16.20

17. $22.50

18. $28\frac{1}{2}$ gallons

19. $2160

20. $33\frac{3}{4}$ yards

Chapter 6

PERCENT

6.1 Basics of Percent

Objective 1

1. 43%   2. 8%   3. 45%   4. 29%

5. 32%   6. 38%   7. 52%

Objective 2

8. 0.37   9. 0.42   10. 0.83   11. 3.10 or 3.1

12. 5.10 or 5.1   13. 0.09   14. 0.04   15. 0.10 or 0.1

16. 0.325   17. 0.619   18. 0.00025   19. 0.00256

Objective 3

20. 30%   21. 40%   22. 20%   23. 90%

24. 71%   25. 86%   26. 42%   27. 7%

28. 9%   29. 3.6%   30. 98.6%   31. 56.4%

32. 493%   33. 347%   34. 420%

Objective 4

35. $19   36. 340 miles   37. $228   38. 12 dogs

Objective 5

39. 24 copies   40. 492 televisions   41. 4 homes   42. 125 signs

6.1 Mixed Exercises

43. 6.75%   44. 26%   45. 0.00302   46. 0.003   47. 0.005

48. 340%   49. 4.23%   50. 7.36%   51. 483.6%   52. 0.05%

53. $250   54. 87 days   55. 49 hours   56. 10 years

6.2 Percents and Fractions

Objective 1

1. $\frac{7}{20}$

2. $\frac{3}{25}$

3. $\frac{14}{25}$

4. $\frac{3}{4}$

5. $\frac{5}{8}$

6. $\frac{109}{250}$

7. $\frac{49}{300}$

8. $\frac{2}{9}$

9. $\frac{1}{15}$

10. $\frac{139}{300}$

Objective 2

11. 70%

12. 53%

13. 81%

14. 48%

15. 85.3%

16. 66%

17. 94%

18. 55.6%

19. 57.1%

20. 380%

21. 275%

22. 740%

Objective 3

23. 0.5, 50%

24. $\frac{1}{8}$, 12.5%

25. 0.25, 25%

26. 0.625, 62.5%

27. $\frac{7}{8}$, 0.875

28. 0.375, 37.5%

29. $\frac{1}{3}$, 0.333

30. 0.4, 40%

31. $\frac{13}{40}$, 32.5%

32. 0.667, 66.7%

6.2 Mixed Exercises

33. $\frac{1}{200}$

34. $\frac{9}{1000}$

35. $\frac{7}{5}$ or $1\frac{2}{5}$

36. $\frac{7}{4}$ or $1\frac{3}{4}$

37. $\frac{9}{4}$ or $2\frac{1}{4}$

38. $\frac{1}{10,000}$

39. 375%

40. 433.3%

41. $\frac{5}{6}$, 0.833

42. 0.714, 71.4%

6.3 Using the Percent Proportion and Identifying the Components in a Percent Problem

Objective 2

1. whole = 120	2. whole = 800	3. whole = 17.5	4. whole = 12
5. part = 12	6. part = 5.4	7. part = 3.5	8. part = 8.5
9. percent = 22	10. percent = 5	11. percent = 26	12. percent = 24

Objective 3

13. 35%	14. 25%	15. 71%	16. 83%
17. 72%	18. Unknown	19. Unknown	20. 42%
21. Unknown	22. 17%		

Objective 4

23. 48	24. 4	25. 78	26. 965
27. Unknown	28. 384	29. 60	30. 487
31. 30	32. Unknown		

Objective 5

33. 560	34. 29	35. 29.81	36. 85
37. Unknown	38. 16.74	39. 4	40. Unknown
41. 720	42. Unknown		

6.3 Mixed Exercises

43. whole = 500	44. part = 66	45. percent = 20	46. 16%
47. 150%	48. 90%	49. 30%	50. 3
51. 200	52. 40	53. 2.4	54. 72
55. 780			

6.4 Using Proportions to Solve Percent Problems

Objective 1

1. 146
2. 280
3. 35.1
4. 3.8

5. 47.3
6. 87.5
7. 118.2
8. 292.5

9. 66 children
10. $84
11. 946 drivers
12. $210

Objective 2

13. 500
14. 300
15. 205.6
16. 152

17. 400
18. 240
19. 498.2
20. 70

21. 7800 students
22. $25,000
23. 75 members
24. 800 applicants

25. 2325 customers
26. $3500

Objective 3

27. 33.3%
28. 26%
29. 16.7%
30. 55%

31. 3%
32. 2.5%
33. 0.2%
34. 0.05%

35. 150%
36. 1125%
37. 2.5%
38. 2.3%

39. 65.5%
40. 1.5%
41. 22%
42. 17%

6.4 Mixed Exercises

43. 1.5
44. 25,000
45. 26%

46. 14.7
47. 13,560
48. 115.4%

49. 1500
50. 3276
51. 3000

52. 5000%
53. 24.5
54. 3000%

55. 150 students
56. 10,600 eligible people
57. 13%

58. 85

6.5 Using the Percent Equation

Objective 1

1. 192
2. 644
3. 845
4. 297
5. 83.2

6. 211.2
7. 21.6
8. 22.5
9. 95
10. 106.4

11. 1004.4
12. 156.1
13. 14 clients
14. $1311.42

Objective 2

15. 400
16. 160
17. 94
18. 1500

19. 1800
20. 850
21. 2160
22. 1200

23. 900
24. 600
25. 15
26. 860

27. 640 gallons
28. 120 units

Objective 3

29. 50%
30. 20%
31. 52%
32. 18%
33. 35%

34. 45%
35. 1.5%
36. 5%
37. 32%
38. 44%

39. 225%
40. 166.7%
41. 25%
42. 29%

6.5 Mixed Exercises

43. 1.4
44. 32
45. 300%
46. 22.9

47. 3.1
48. 333.3%
49. 2000
50. 1.0

51. 55.6%
52. 12
53. 245%
54. 2109.5

6.6 Solving Application Problems

Objective 1

1. $3; $103
2. $12; $212
3. $3.50; $53.50
4. $15.30; $185.30
5. $10.75; $225.75
6. $0.30; $15.30
7. $2.40; $32.40
8. $0.78; $78.78
9. $6.03; $73.03
10. $31.50; $481.50
11. $22.75
12. $931.25

Objective 2

13. $15
14. $40
15. $110
16. $231.32
17. $1024
18. $225
19. $155.63
20. $3000
21. $4680
22. $3750

Objective 3

23. $25; $75
24. $30; $170
25. $78; $702
26. $15.20; $22.80
27. $8.75; $8.75
28. $6.75; $15.75
29. $43.75; $81.25
30. $14.97; $9.98
31. $119.16; $476.64
32. $10.28; $195.22

Objective 4

33. 20%
34. 25%
35. 27.1%
36. 28.3%
37. 131.3%
38. 44.0%
39. 8.3%
40. 28%

6.6 Mixed Exercises

41. $1001.70
42. $810
43. 5.5%
44. $24
45. $1830.60
46. 6%
47. 2.5%
48. $4455
49. $36.18
50. $1408.64
51. $9.75
52. $836.50
53. $12,884.25
54. $112.50; $337.50
55. 6.5%
56. 441.7%

.7 Simple Interest

Objective 1

1. $20
2. $24
3. $144
4. $240
5. $4

6. $45.50
7. $270
8. $2112
9. $8
10. $18

11. $41.25
12. $120
13. $36.90
14. $6.90
15. $19.50

16. $31.32
17. $95.20
18. $120

Objective 2

19. $222
20. $3075
21. $556.20
22. $1224
23. $1680

24. $5145.83
25. $2309.45
26. $6169.20
27. $1230
28. $9240

.7 Mixed Exercises

29. $195
30. $32.40
31. $124
32. $780

33. $29.25
34. $173.60
35. $592.08
36. $360

37. $1886.80
38. $27,720
39. $19,049.33
40. $23,408

41. $1170
42. $292.50
43. $1905
44. $2149

6.8   Compound Interest

Objective 3

1. $8103.38
2. $3401.22
3. $1273.45
4. $3663.68
5. $4867.20
6. $8337.11

Objective 4

7. $1262.50
8. $1850.90
9. $6205.20
10. $11,277
11. $66.78
12. ≈ $60.47
13. $47.24
14. $91.83
15. $10,976.01
16. $19,815.73

Objective 5

17. $1628.90; $628.90
18. $1272.30; $272.30
19. $17,103.70; $8603.70
20. $20,724.48; $7924.48
21. $16,935.74; $7785.74
22. $52,645.50; $7645.50
23. $34,730.06; $13,330.06
24. $111,212.40; $33,212.40

6.8   Mixed Exercises

25. $20,145.60
26. $4282.24
27. $1302.30
28. $29,901.30
29. $42,068.14
30. $14,282.10; $5282.10
31. $3345.50; $845.50

MEASUREMENT

.1 Problem Solving with English Measurement

Objective 1

1. 12
2. 16
3. 2
4. 2000
5. 5280
6. 4
7. 8
8. 3
9. 24
10. 1

Objective 2

11. 6 ft
12. 1 gal
13. 3 c
14. 48 hr
15. 2 mi
16. 21 ft
17. 12 pt
18. 10 qt
19. 150 min
20. 15,840 ft

Objective 3

21. 4 yd
22. $3\frac{1}{2}$ gal
23. 3 gal
24. $3\frac{3}{4}$ lb
25. $1\frac{1}{2}$ T
26. 21,120 ft
27. $\frac{5}{6}$ yd
28. 19 pt
29. 120 hr
30. 180 min

Objective 4

31. 10,000 lb
32. (a) 3.5 T   (b) 1274 T

.1 Mixed Exercises

33. $5.31
34. 5 gal
35. 56 oz
36. 13 qt
37. 210 in
38. $7\frac{1}{2}$ ft
39. $8\frac{4}{7}$ wk
40. $5\frac{1}{4}$ mi
41. $1\frac{1}{4}$ min
42. $6\frac{1}{3}$ hr
43. $6.36
44. 22.5 qt

7.2 The Metric System – Length

Objective 2

1. 1000 mm 2. 0.001 km 3. 100 cm 4. 0.001 m 5. 0.01 m

6. 1000 m 7. 7000 mm 8. 2587 cm 9. 2300 mm 10. 5310 cm

Objective 3

11. 0.636 m 12. 0.807 m 13. 96 m 14. 0.14 m 15. 2297 mm

16. 19.4 mm 17. 6400 m 18. 10,350 m 19. 14.5 km 20. 25.693 km

7.2 Mixed Exercises

21. 350,000 cm 22. 286,000 cm 23. 720 mm 24. 60.2 cm

25. 610 m 26. 0.0836 km 27. More 28. More

29. More 30. More

.3 The Metric System – Capacity and Weight (Mass)

Objective 2

1. 0.007 kL
2. 9700 mL
3. 2500 mL
4. 32,400,000 mL

5. 836,000 L
6. 0.523 L
7. 7.863 L
8. 0.007724 kL

Objective 4

9. 9 kg
10. 27 kg
11. 6300 g
12. 760 g

13. 4700 mg
14. 4,910,000 mg
15. 0.008745 kg
16. 0.042 g

Objective 5

17. mg
18. L
19. km
20. cm

7.3 Mixed Exercises

21. 9750 g
22. 0.00871 kL
23. Unreasonable
24. Reasonable

25. Unreasonable
26. Reasonable
27. mL
28. m

7.4 Problem Solving with Metric Measurement

Objective 1

1. 60 servings
2. $2.61
3. $12.99

4. 8.925 kg
5. 500 bottles
6. 5600 kg

7. 8.33 m
8. 1.935 kg
9. 2.45 L

10. 200 g

7.5 Metric–English Conversion and Temperature

Objective 1

1. 32.7 yd
2. 42.3 ft
3. 74.8 in
4. 8.8 mi

5. 21.3 lb
6. 49.6 qt
7. 45.5 m
8. 3.6 m

9. 527.8 m
10. 4.7 m
11. 18.1 kg
12. 55.2 L

13. $4.96
14. 2.7 qt
15. 73.4 kg
16. 468.5 km

Objective 3

17. 17° C
18. 35° C
19. 45° C
20. 230° C
21. 27° C

22. 52° C
23. 50° F
24. 86° F
25. 32° F
26. 140° F

27. 212° F
28. 302° F
29. 46° C
30. 392° F

7.5 Mixed Exercises

31. 53.4 qt
32. 1.9 lb
33. 5.2 gal
34. 67.0 mi

35. 11.4 L
36. 2.4 m
37. 115.9 km
38. 78.8 kg

39. $0.34
40. $11.44
41. $2.74
42. Gallon bottle at $3.60

43. 842° F
44. 4° C
45. 91° F
46. 11° C

Chapter 8

GEOMETRY

8.1 Basic Geometric Terms

Objective 1

1. Line segment: $\overline{CD}$ or $\overline{DC}$

2. Ray: $\overrightarrow{BA}$

3. Line: $\overleftrightarrow{PQ}$ or $\overleftrightarrow{QP}$

4. Line: $\overleftrightarrow{EF}$ or $\overleftrightarrow{FE}$

5. Line segment: $\overline{KL}$ or $\overline{LK}$

6. Ray: $\overrightarrow{RS}$

7. Line: $\overleftrightarrow{ST}$ or $\overleftrightarrow{TS}$

8. Line segment: $\overline{MN}$ or $\overline{NM}$

9. Line: $\overleftrightarrow{VW}$ or $\overleftrightarrow{WV}$

10. Ray: $\overrightarrow{CB}$

Objective 2

11. Parallel

12. Parallel

13. Intersecting

14. Parallel

15. Intersecting

16. Parallel

17. Parallel

18. Intersecting

19. Parallel

20. Intersecting

Objective 3

21. $\angle COD$ or $\angle DOC$

22. $\angle MON$ or $\angle NOM$

23. $\angle ROS$ or $\angle SOR$

24. $\angle DAE$ or $\angle EAD$

25. $\angle WQR$ or $\angle RQW$

26. $\angle NLM$ or $\angle MLN$

27. $\angle HEI$ or $\angle IEH$

28. $\angle VSW$ or $\angle WSV$

29. $\angle ZRM$ or $\angle MRZ$

30. $\angle BEC$ or $\angle CEB$

Objective 4

31. Obtuse

32. Acute

33. Straight

34. Right

35. Straight

36. Acute

37. Right

38. Obtuse

39. Obtuse

40. Straight

Objective 5

41. Perpendicular

42. Parallel

43. Intersecting

44. Perpendicular   45. Intersecting   46. Perpendicular

47. Intersecting   48. Parallel   49. Intersecting

50. Perpendicular

8.1 Mixed Exercises

51. Ray: $\overrightarrow{RS}$ 52. Line: $\overleftrightarrow{TU}$ or $\overleftrightarrow{UT}$ 53. Line segment: $\overline{WX}$ or $\overline{XW}$

54. Ray: $\overrightarrow{YZ}$ 55. Intersecting   56. Parallel

57. Parallel   58. Intersecting   59. $\angle TSV$ or $\angle VST$

60. $\angle EFI$ or $\angle IFE$ 61. $\angle NKL$ or $\angle LKN$ 62. $\angle OPS$ or $\angle SPO$

63. Acute   64. Straight   65. Right

66. Obtuse   67. Parallel   68. Perpendicular

69. Intersecting   70. Parallel

.2 Angles and Their Relationships

)bjective 1

1. 78°   2. 47°   3. 18°   4. 24°

5. 59°   6. 164°   7. 12°   8. 142°

9. ∠BAC and ∠CAD;   10. ∠SRT and ∠TRU;
    ∠DAE and ∠EAF   ∠URV and ∠VRW

11. ∠LKM and ∠MKN;   12. ∠QPR and ∠RPS;
     ∠NKO and ∠OKL;   ∠SPT and ∠TPQ;
     ∠MKN and ∠NKO;   ∠RPS and ∠SPT;
     ∠OKL and ∠LKM   ∠TPQ and ∠QPR

)bjective 2

13. ∠MON and ∠POQ; ∠NOP and ∠QOM

14. ∠YXZ and ∠ZXW and ∠WXV and ∠VXY

15. ∠ACB and ∠DCE; ∠ACD and ∠BCE

16. ∠MKQ and ∠NKP; ∠MKN and ∠QKP

17. ∠CAD and ∠BAE

18. ∠GFH and ∠HFJ

19. 42°   20. 105°   21. 33°   22. 33°

.2 Mixed Exercises

23. 63°   24. 28°   25. 1°   26. 75°   27. 48°

28. 84°   29. 90°   30. 135°   31. 113°   32. 38°

33. 80°   34. 58°   35. 44°   36. 5°   37. 50°

38. 93°   39. 37°   40. 50°

8.3 Rectangles and Squares

Objective 1

1.   24 cm; 32 cm^2

2.   58 in; 204 in^2

3.   36 cm; 17 cm^2

4.   35.4 m; 46.4 m^2

5.   22 yd; $29\frac{1}{4}$ yd^2

6.   240.4 ft; 2877.6 ft^2

7.   281.2 cm; 3859.68 cm^2

8.   100 in; 600 in^2

9.   7248 ft^2

10. $5796

Objective 2

11. 36 m; 81 m^2

12. 36.8 yd; 84.64 yd^2

13. 31.2 ft; 60.84 ft^2

14. 52 ft; 169 ft^2

15. $5\frac{3}{5}$ in; $1\frac{24}{25}$ in^2

16. 32.8 km; 67.24 km^2

17. 12.4 cm; 9.61 cm^2

18. 29.6 in; 54.76 in^2

19. $18\frac{2}{3}$ mi; $21\frac{7}{9}$ mi^2

20. 84 m; 4441 m^2

Objective 3

21. 30 ft; 18 ft^2

22. 48 in; 83 in^2

23. 30 m; 30 m^2

24. 42 yd; 50 yd^2

25. 32 in; 28 in^2

26. 24 cm; 31 cm^2

27. 76 mm; 192 mm^2

28. 48 ft; 101 ft^2

29. 42 yd; 54 yd^2

30. 32 km; 46 km^2

8.3 Mixed Exercises

31. 18 m; 20 m^2

32. 44 yd; 117 yd^2

33. 36 ft; 65 ft^2

34. 15.6 m; 15.21 m^2

35. 248 cm; 3844 cm^2

36. $1\frac{1}{3}$ in; $\frac{1}{9}$ in^2

37. 46 m; 65 m^2

38. 52 ft; 87 ft^2

39. 98 m; 460 m^2

40. 94 cm; 260 cm^2

41. 56 cm; 171 cm^2

42. 188 ft; 1348 ft^2

8.4 Parallelograms and Trapezoids

Objective 1

1. 168 m

2. 32.8 in

3. 30 ft

4. 713 yd^2

5. 123.48 m^2

6. $11\frac{1}{4}$ m^2

7. 14.72 m^2

8. 310 ft^2

9. $587.52

10. $780

Objective 2

11. 159 in

12. 708.8 cm

13. $53\frac{3}{4}$ yd

14. 1106 m^2

15. 1943.7 cm^2

16. 60 in^2

17. $43\frac{3}{4}$ ft^2

18. $9\frac{5}{8}$ in^2

19. 32.5 cm^2

20. 4190 cm^2

8.4 Mixed Exercises

21. 28 m

22. 26.0 in

23. 22 cm

24. 19.22 yd^2

25. 27 in^2

26. $34\frac{7}{8}$ m^2

27. 3515.4 cm^2

28. 15,504 ft^2

29. $7579

30. $700

31. $174.60

8.5   Triangles

Objective 1

1. 25 yd
2. 59 cm
3. 34 yd
4. 5 m

5. 37.2 ft
6. 3 in
7. $10\frac{3}{4}$ ft
8. 11.3 cm

9. 17.7 m
10. $39\frac{3}{8}$ in

Objective 2

11. 1260 m^2
12. 232.56 cm^2
13. $21\frac{3}{4}$ ft^2
14. 28 yd^2

15. 15.81 m^2
16. $\frac{77}{128}$ in^2
17. 42 ft^2
18. 510 m^2

19. 1008 m^2
20. 534 cm^2

Objective 3

21. 50°
22. 17°
23. 50°
24. 60°
25. 20°

26. 81°
27. 59°
28. 57°
29. 90°
30. 45°

8.5   Mixed Exercises

31. 30 m
32. 13.1 in
33. $24\frac{1}{2}$ ft
34. 1609.35 m^2

35. 1940 yd^2
36. $190.82
37. 40°
38. 40°

39. 28°
40. 61°

8.6 Circles

Objective 1

1. 86 m   2. 128 ft   3. 13.25 m   4. 29.5 in   5. 5.8 ft

6. 7 m   7. 6.3 in   8. 6.65 m   9. $4\frac{7}{8}$ in   10. $\frac{1}{4}$ in

Objective 2

11. 69.1 ft   12. 144.4 cm   13. 94.2 m   14. 40.8 in

15. 106.8 ft   16. 28.3 yd   17. 62.8 cm   18. 14.9 in

19. 4.7 mi   20. 40.1 cm

Objective 3

21. 78.5 in^2 22. 43.0 m^2 23. 1519.8 yd^2 24. 2041.8 ft^2

25. 22.3 yd^2 26. 75.4 cm^2 27. 57 cm^2 28. 248.5 m^2

29. 2101.3 m^2 30. 298.3 cm^2

8.6 Mixed Exercises

31. 4 ft   32. 5.4 cm   33. $6\frac{1}{4}$ yd   34. 188.4 cm

35. 18.8 m   36. 18.2 ft^2 37. 28.8 m^2 38. 124.6 m^2

39. $320.87   40. 241.4 ft^2

8.7 Volume

Objective 1

1. 468 cm^3
2. 96 ft^3
3. 2744 in^3
4. 360 in^3

5. 400 m^3
6. 176 in^3
7. 273 cm^3
8. 729 km^3

9. 95.2 cm^3
10. 624 cm^3

Objective 2

11. 4.2 m^3
12. 0.5 m^3
13. 1071.8 in^3
14. 2.8 in^3

15. 16.7 cm^3
16. 113.0 ft^3
17. 7.1 km^3
18. 1562.7 in^3

19. 2.6 cm^3
20. 18.0 in^3

Objective 3

21. 471 ft^3
22. 75.4 m^3
23. 942 ft^3
24. 1846.3 in^3

25. 141.3 ft^3
26. 0.1 km^3
27. 508.7 m^3
28. 602.9 ft^3

29. 1433.5 in^3
30. 1105.3 in^3

Objective 4

31. 121.3 m^3
32. 1520 cm^3
33. 452.2 m^3
34. 7536 cm^3

35. 6 km^3
36. 33.5 m^3
37. 150.7 yd^3
38. 324 cm^3

39. $22{,}344 \text{ m}^3$
40. 15.5 cm^3

8.7 Mixed Exercises

41. 48 m^3
42. 2310 m^3
43. 65.4 in^3
44. 3267.5 ft^3

45. 1808.6 cm^3
46. 257.2 cm^3
47. 678.2 cm^3
48. 490.6 cm^3

49. 25.8 ft^3
50. 339.1 ft^3

8.8 Pythagorean Theorem

Objective 1

1. 4.123	2. 5.196	3. 7.874	4. 7.416
5. 4.899	6. 3.162	7. 1.414	8. 6.083
9. 3.606	10. 5.292	11. 6.856	12. 7.280
13. 8.426	14. 8.660	15. 10.100	16. 12.042

Objective 2

17. 2.646 in	18. 1.414 cm	19. 9.849 in	20. 2.236 ft
21. 2.236 cm	22. 12.104 in	23. 10.198 m	24. 5.745 yd
25. 13.229 km	26. 2.907 m	27. 1.008 in	28. 8.660 cm

Objective 3

29. 8.062 ft	30. 3.162 yd	31. 18.028 mi	32. 35.199 ft
33. 8 ft	34. 24 ft	35. 467.039 cm	36. 10 ft
37. 7.616 cm	38. 8.485 cm		

8.8 Mixed Exercises

39. 13.784	40. 15.811	41. 12.369	42. 13.601
43. 5 cm	44. 13 ft	45. 8.5 in	46. 7.5 m
47. 12 km	48. 12.649 cm	49. 9.451 in	50. 3.612 ft
51. 4.031 yd	52. 8.602 in	53. 4.472 in	

8.9 Similar Triangles

Objective 1

1. $\angle A$ and $\angle P$, $\angle C$ and $\angle R$, $\angle B$ and $\angle Q$; $\overline{AB}$ and $\overline{PQ}$, $\overline{AC}$ and $\overline{PR}$, $\overline{BC}$ and $\overline{QR}$

2. $\angle C$ and $\angle R$, $\angle B$ and $\angle Q$, $\angle A$ and $\angle P$; $\overline{AB}$ and $\overline{PQ}$, $\overline{AC}$ and $\overline{PR}$, $\overline{BC}$ and $\overline{QR}$

3. $\angle M$ and $\angle Q$, $\angle N$ and $\angle R$, $\angle P$ and $\angle S$; $\overline{MP}$ and $\overline{QS}$, $\overline{MN}$, and $\overline{QR}$, $\overline{NP}$ and $\overline{RS}$

4. $\angle X$ and $\angle P$, $\angle Y$ and $\angle Q$, $\angle Z$ and $\angle R$; $\overline{XY}$ and $\overline{PQ}$, $\overline{XZ}$ and $\overline{PR}$, $\overline{YZ}$ and $\overline{QR}$

5. $\angle G$ and $\angle S$, $\angle H$ and $\angle R$, $\angle K$ and $\angle T$; $\overline{GH}$ and $\overline{SR}$, $\overline{GK}$ and $\overline{ST}$, $\overline{KH}$ and $\overline{TR}$

6. $\angle J$ and $\angle P$, $\angle K$ and $\angle M$, $\angle L$ and $\angle N$; $\overline{JL}$ and $\overline{PN}$, $\overline{JK}$ and $\overline{PM}$, $\overline{KL}$ and $\overline{MN}$

7. $\angle D$ and $\angle W$, $\angle F$ and $\angle X$, $\angle G$ and $\angle Y$; $\overline{DF}$ and $\overline{WX}$, $\overline{DG}$ and $\overline{WY}$, $\overline{GF}$ and $\overline{YX}$

Objective 2

8. $a = 15$, $b = 9$

9. $a = 6.7$, $b = 20$

10. $a = 6$, $b = 8$

11. 3

12. 10

13. 18

14. 60

15. 90

16. 6.6

17. 9

Objective 3

18. 21 ft

19. 36 m

20. 132 yd

21. 24.3 m

22. 45 m, 60 m

23. 6.4 cm, 8 cm

24. 56

25. 10

26. 80

27. 108

8.9 Mixed Exercises

28. $\angle A$ and $\angle R$, $\angle B$ and $\angle S$, $\angle C$ and $\angle T$; $\overline{AB}$ and $\overline{RS}$, $\overline{BC}$ and $\overline{ST}$, $\overline{CA}$ and $\overline{TR}$

29. $\angle X$ and $\angle N$, $\angle Y$ and $\angle O$, $\angle Z$ and $\angle M$; $\overline{XY}$ and $\overline{NO}$, $\overline{YZ}$ and $\overline{OM}$, $\overline{XZ}$ and $\overline{NM}$

30. $\angle J$ and $\angle L$, $\angle H$ and $\angle M$, $\angle I$ and $\angle K$; $\overline{HI}$ and $\overline{MK}$, $\overline{JI}$ and $\overline{LK}$, $\overline{JH}$ and $\overline{LM}$

31. 22   32. 13.3   33. 95   34. 16

35. 24   36. 31.3   37. 7.2

Chapter 9

BASIC ALGEBRA

9.1 Signed Numbers

Objective 1

1. +17

2. −120

3. +830

4. −15

5. −13

6. +65

7. −120

8. +14,000

9. +32,000

10. −185

11. Positive

12. Neither

13. Negative

14. Negative

15. Negative

16. Positive

17. Positive

18. Negative

19. Negative

20. Positive

Objective 2

21.

22.

23.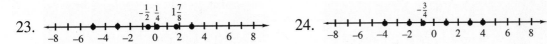

24.

25.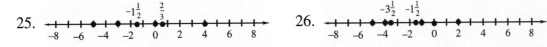

26.

27.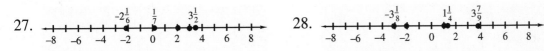

28.

29.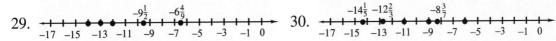

30.

Objective 3

31. <

32. <

33. >

34. >

35. <

36. <

37. <

38. >

39. <

40. <

41. <

42. >

Objective 4

43. 6

44. 11

45. −8

46. −352

47. 0

48. $\frac{10}{7}$ 49. −8.23   50. 5.6   51. 7.2

52. 8.7   53. $-\frac{1}{3}$ 54. $\frac{8}{9}$

Objective 5

55. −3   56. −8   57. 3   58. 10

59. −12   60. 72   61. 43   62. −41

63. $-\frac{5}{9}$ 64. $-\frac{9}{10}$ 65. $\frac{2}{3}$ 66. $\frac{7}{10}$

9.1 Mixed Exercises

67. −7   68. −30

69. +12   70. Negative

71. Positive   72.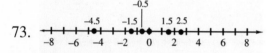

73.   74. >

75. >   76. >

77. $\frac{5}{7}$ 78. $\frac{11}{3}$

79. −7   80. −13

81. −3   82. −26

83. −6.1   84. −4.8

85. 2.5

9.2 Adding and Subtracting Signed Numbers

Objective 1

1. 6	2. 2	3. 2	4. –3	5. –7
6. –3	7. –1	8. –7	9. –9	10. –6

Objective 2

11. 7	12. –5	13. –12	14. –7.9	15. –16.9
16. –2.82	17. 0.83	18. $\frac{3}{4}$	19. $\frac{1}{2}$	20. $-\frac{21}{10}$
21. $-3\frac{1}{4}$	22. $\frac{8}{9}$			

Objective 3

23. –4	24. –5	25. 10	26. 15	27. 8
28. 13	29. –77	30. –281	31. 0	32. –3.5
33. –2.7	34. 4.6			

Objective 4

35. –7	36. –3	37. –55	38. –11	39. 8
40. –12	41. –72	42. –1.2	43. –3.4	44. 12
45. –3	46. 3	47. 4	48. $\frac{5}{4}$	49. $-\frac{1}{8}$

Objective 5

50. –12	51. –10	52. –14	53. 13	54. 25
55. 6	56. $-\frac{1}{3}$	57. 2	58. –11.5	59. 5.2

9.2 Mixed Exercises

60. –6	61. –1	62. –9	63. –69	64. –19
65. 6	66. $-8\frac{1}{2}$	67. $-7\frac{1}{2}$	68. 5.11	69. $-\frac{3}{8}$
70. $-\frac{15}{7}$	71. $\frac{7}{9}$	72. $\frac{3}{5}$	73. $-\frac{4}{11}$	74. –16

75. −17   76. −3   77. −18.1   78. −12.7   79. −8

80. 12   81. −1   82. 3.1   83. 9.5

9.3 Multiplying and Dividing Signed Numbers

Objective 1

1. −21   2. −24   3. −50   4. −54   5. −12

6. −96   7. $-\frac{5}{6}$ 8. $-\frac{3}{4}$ 9. −33.3   10. −32.2

11. −26.98   12. −29.76   13. −2   14. −4   15. −4

16. −8   17. $-\frac{5}{4}$ 18. −11

Objective 2

19. 44   20. 90   21. 42   22. 18   23. $\frac{5}{2}$

24. $\frac{12}{5}$ 25. 13   26. 2   27. 8   28. 7

29. $\frac{3}{16}$ 30. $\frac{1}{2}$ 31. $\frac{1}{3}$ 32. $\frac{9}{7}$ 33. $\frac{3}{4}$

9.3 Mixed Exercises

34. −80   35. −39   36. −108

37. $-\frac{8}{3}$ 38. −5.92   39. −8

40. $\frac{9}{2}$ 41. $\frac{7}{8}$ 42. $\frac{9}{2}$

43. $\frac{3}{5}$ 44. $\frac{1}{16}$ 45. 4.8

46. 4.55

9.4 Order of Operations

Objective 1

1. −20
2. −9
3. −8
4. 2

5. −10
6. 28
7. 6
8. 30

Objective 2

9. 40
10. 125
11. −3
12. 3
13. −25

14. 45
15. $-\frac{38}{63}$
16. $-\frac{1}{12}$
17. −6
18. $\frac{102}{7}$

Objective 3

19. −1
20. −2
21. −1
22. −1

9.4 Mixed Exercises

23. 14
24. −80
25. −7
26. −12

27. −3
28. −3
29. 31
30. 24

31. −2
32. −12
33. −18
34. $\frac{5}{4}$

35. $-\frac{14}{25}$
36. $-\frac{26}{25}$
37. −2
38. 2

9.5 Evaluating Expressions and formulas

Objective 2

1. -4 2. -11 3. 24 4. -17 5. -15

6. 1 7. 24 8. -12 9. -21 10. -4

11. 2 12. 1 13. $P = 20$ 14. $P = 20$ 15. $P = 28$

16. $P = 52$ 17. $A = 36$ 18. $A = 39$ 19. $V = 50$ 20. $V = 68$

21. $d = 195$ 22. $d = 880$ 23. $C \approx 56.52$ 24. $C \approx 75.36$

9.6 Solving Equations

Objective 1

1. Yes   2. Yes   3. No   4. No

5. No   6. Yes   7. No   8. Yes

9. No   10. Yes   11. Yes   12. No

Objective 2

13. 4   14. 10   15. 14   16. 5   17. 11   18. 5   19. −2

20. 13   21. 1   22. −2   23. −4   24. −2   25. $\frac{2}{3}$ 26. $\frac{19}{8}$

Objective 3

27. 4   28. 7   29. −6   30. −6

31. 7   32. 16   33. 32   34. −14

35. −50   36. −21   37. −36   38. 12

9.6 Mixed Exercises

39. No   40. Yes   41. Yes   42. No

43. $\frac{47}{4}$ 44. $\frac{20}{3}$ 45. $\frac{17}{3}$ 46. $\frac{41}{9}$

47. 5.61   48. 10.26   49. −2.61   50. 7

51. −1.1   52. 2.6   53. −32   54. 20

55. $\frac{5}{12}$ 56. $\frac{3}{8}$ 57. 2.1   58. 5.72

9.7 Solving Equations with Several Steps

Objective 1

1. 1
2. 2
3. 11
4. 2
5. 2

6. 3
7. –3
8. –4
9. –3
10. –3

Objective 2

11. 56
12. 153
13. $3x+21$
14. $7k-35$

15. $-18m-3m$
16. $-10-5a$
17. $-2y+6$
18. $-5-a$

19. $-15+3a$
20. $5-x$

Objective 3

21. $16r$
22. $2m$
23. $16x$
24. $18y$
25. $-36m$

26. $5z$
27. $-8k$
28. $-5a$
29. $-7z$
30. $-1.9x$

Objective 4

31. 5
32. 6
33. 5
34. 7
35. 5

36. 8
37. 2
38. 3
39. –1
40. 1

41. 20
42. –18
43. –2
44. –7
45. 8

46. 4
47. 0
48. –11
49. –1.8
50. –1

51. $\frac{5}{2}$
52. 1

9.7 Mixed Exercises

53. $-35+5r$
54. $-32+4x$
55. $2.3a$
56. $-\frac{1}{2}k$
57. 6

58. 16
59. 4.8
60. –2.7
61. –2
62. –7

63. 7
64. –3
65. –6
66. 1
67. –2

68. –3
69. 3
70. 8
71. $\frac{7}{3}$
72. 65

9.8   Using Equations to Solve Applications Problems

Objective 1

1. $13+x$
2. $9+x$
3. $5+x$
4. $x+(-9)$

5. $x-5$
6. $x-6$
7. $3x$
8. $2x$

9. $2x+7$
10. $3x+3$
11. $3x+5x$
12. $10x-6x$

Objective 2

13. $-5n+4=-11;\ 3$
14. $45-2n=35;\ 5$
15. $3n+7=1;\ -2$

16. $6+4n=50;\ 11$
17. $2n-3=-17;\ -7$
18. $2n+18=4n;\ 9$

Objective 3

19. 4
20. 3
21. 7
22. 5

23. 22
24. 15
25. –10
26. –3

27. 38 cm
28. 247 votes
29. 5 days

9.8   Mixed Exercises

30. $x+7$
31. $x+3$
32. $x-8$

33. $1-x$
34. $3x$
35. $\frac{x}{2}$ or $\frac{1}{2}x$

36. $\frac{x}{4}$
37. $\frac{7}{x}$
38. $3n-12=3;\ 5$

39. $9n-7n=16;\ 8$
40. $8-2n=20+n;\ -4$
41. $\frac{1}{2}n+4=10;\ 12$

42. 250 mi
43. 14 m
44. 8 cm

45. 223 m
46. 21 m
47. 5 years

48. $7000

Chapter 10

STATISTICS

10.1 Circle Graphs

Objective 1

1. 896 students
2. 640 students
3. 608 students
4. 576 students

5. 288 students
6. 192 students
7. Liberal Arts
8. Computer Science

9. 3:1
10. 5:7

Objective 2

11. $10,400
12. Carpentry
13. $\frac{3}{8}$
14. $\frac{7}{104}$

15. $\frac{3}{14}$
16. $\frac{1}{2}$
17. Business
18. $\frac{10}{29}$

19. $\frac{9}{58}$
20. $\frac{5}{9}$
21. $\frac{1}{4}$
22. $\frac{9}{20}$

23. $285,000
24. $237,500
25. $95,000
26. $95,000

27. $142,500
28. $95,000

Objective 3

29. 20%

30. 15%, 54°

31. 36°

32. 15%

33. 5%, 18°

34. $160, 10%

35.

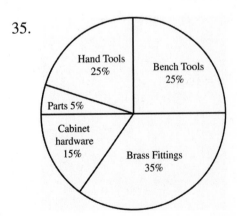

10.1 Mixed Exercises

36. $8560
37. $14,980
38. $5564

39. $5136
40. $4280
41. $4280

42. (a) $400,000

(b) Parts: 5%, hand tools: 20%, bench tools: 25%, brass fittings: 35%, cabinet hardware: 15%

(c) Parts: 18°, hand tools: 72°, bench tools: 90°, brass fittings: 126°, cabinet hardware: 54°

(d)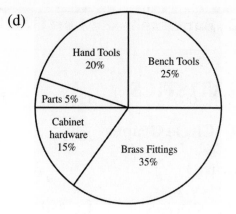

43. (a) Mysteries: 108°, biographies: 54°, cookbooks: 36°, romantic novels: 90° science: 54°, business: 18°

(b)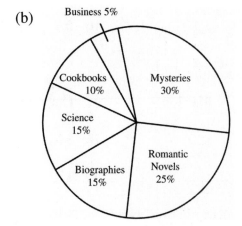

44. (a) Housing: 144°, food: 72°, automobile: 50.4^o, clothing: 28.8^o, medical: 21.6^o, savings: 28.8^o, other 14.4^o

(b)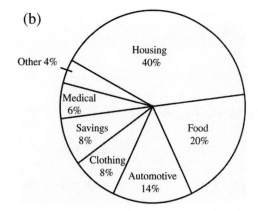

10.2 Bar Graphs and Line Graphs

Objective 1

1. 1000   2. 1300   3. 1600   4. 1400   5. 1800

6. 300   7. 2002   8. 2001   9. 300   10. 2002

Objective 2

11. Sophomore   12. 350   13. 400   14. 650

15. 850   16. 10:7   17. 9:5   18. 2900

19. 17:11   20. Freshman

Objective 3

21. September   22. October   23. $20   24. $30

25. $20   26. $40   27. 3:1   28. Increased

29. $30   30. $30

Objective 4

31. $3,500,000   32. $1,000,000   33. $1,500,000

34. $2,500,000   35. $2,000,000   36. $3,000,000

37. 2000, 2001, and 2002   38. 2000   39. 1998

40. 1:2

10.2 Mixed Exercises

41. 3600   42. July 3   43. 2400   44. 4200

45. July 4   46. 3000   47. 1800   48. 600

49. July 5 and 6   50. 16,800   51. April; 900   52. 300

53. 300   54. 200   55. 400   56. 200

57. 600,000   58. 200,000   59. 1999   60. 2000

61. 200,000   62. 200,000   63. June   64. February

65. 500   66. 200   67. 200   68. 300

69. 100

70. Increase

71. 400 more burglaries

72. 2600

73. $15,000

74. $30,000

75. $40,000

76. $5000

77. $10,000

78. $20,000

10.3 Frequency Distributions and Histograms

Objective 1

1. ||; 2

2. |||; 3

3. |||| |; 6

4. ||; 2

5. ||||; 4

6. |; 1

Objective 2

7. |||| ; 5

8. |||| |||; 8

9. |||; 3

10. |; 1

11. |||; 3

12. |; 1

13. 110–129

14. 150–169 and 190–209

Objective 3

15. 61–65

16. 16–20

17. 14 members

18. 120 members

19. 82 members

20. 30 members

10.3 Mixed Exercises

21. |; 1

22. |||| ; 5

23. |||| ||; 7

24. ||||; 4

25. |; 1

26. 100–109

27. 80–89 and 120–129

28.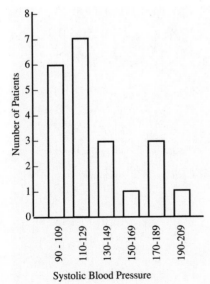

29.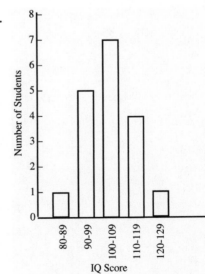

10.4 Mean, Median, and Mode

Objective 1

1. 7.2

2. 51.8

3. 60.3

4. 55.1

5. 6.2

6. 64.5

7. 27,955

8. 40,527

9. 7.7

10. 52.7

Objective 2

11. 6.8

12. 14.8

13. 18.6

14. 40.2

15. 4.7

16. 27.1

17. 1.7

18. 3.1

19. 2.5

20. 2.5

Objective 3

21. 42.5

22. 298

23. 232

24. 29.4

25. .006

26. 1.7

27. 632

28. 57

29. 2067

30. 937

Objective 4

31. 8

32. 43

33. 5

34. 3

35. 79, 85

36. 272

37. No mode

38. No mode

39. 24, 35, and 39

40. 172.6

10.4 Mixed Exercises

41. 12.4

42. 5.4

43. 273.1

44. 16.4

45. 32.2

46. 25

47. 1.8

48. 239.5

49. 8

50. 0.7